Published in the United Kingdom in 2026 by
Midland Tutorial Productions

First Edition 1 June 2026

Cover Design: James Warren

Cover Photograph:
A Duck Family on Ellesmere
Olympus TG5: 09:16GMT 8 April 2026
James R Warren

File Prefix Code: QUAD

ISBN 978 1 915750 17 4

Midland Tutorial Productions
31 Victoria Avenue, Bloxwich, Walsall, WS3 3HS
United Kingdom

SIMPLE QUADRATURE

FOR CLOSED PLANAR SHAPES

James R Warren

BLOXWICH
2026

SIMPLE QUADRATURE

FOR CLOSED PLANAR SHAPES

James R Warren

MIDLAND TUTORIAL PRODUCTIONS
BLOXWICH

Other Books By James R Warren

Boscawen-Ûn
Beyond Tourist Britain
Gleanings as I Pass
Exordium
Meditations
Gamma Solution
Moddeshall Hydropower
Unreasonable Mathematics
Mathematical Explorations
Researches: Volume One
Researches: Volume Two
Researches: Volume Three
Researches: Volume Four
Pi and Phi
Four Famous Numbers
Progress in Iron Smelting
Congleholme
Acts of Trust
TAM
The Mystic Beauty of the Trefoil

To The Glory of The Loving God

Who Made Our Minds Free

ACKNOWLEDGEMENTS

The National Library of Scotland

I gratefully acknowledge the permission of The National Library of Scotland to re-produce images from their map archive as bases for my demonstrations of geomorphological principles and quadratures, as downloaded from their Online Map Archive and in particular https://maps.nls.uk/geo/explore/side-by-side/#zoom=5.0&lat=56.00000&lon=-4.00000&layers=1&right=ESRIWorld

The NLS writes:-
CC-BY. You are welcome to use some images displayed on the Map Images website for commercial, non-commercial, educational and private purposes under the terms of the Creative Commons Attribution (CC-BY) licence.
(https://maps.nls.uk/copyright.html)

Several of these maps originally bore the Crown Copyright as productions of The Ordnance Survey, but to the best of my believe none post-date 1971AD and as corporate productions are therefore copyright-expired in the United Kingdom.

The Brathay Exploration Group

I gratefully acknowledge the provision of several bathymetric maps of Cumbrian tarns re-produced as bases for my demonstrations of geomorphological principles and quadratures.

These images were digitally copied from "*Tarns of the Central Lake District*", a BEG publication cited in the Bibliography of this book.

TABLE OF CONTENTS

Contents

PART ONE
PRINCIPLES AND PRELIMINARIES

Quadrature is the mathematical art of estimating the areas of surfaces, especially plane (i.e. flat) surfaces.

The mathematicians tell us the triangle is the most primitive of those closed plane figures that have straight-line sides. A triangle has three sides.

They further advise that any triangle has an infinity of "centers". Like many things in mathematics and science I find this entirely counter-intuitive. Except where there are special circumstances, I prefer to think in physical terms and choose a center accordingly. Of course a triangle is not a physical object. Though you may show its form with a pencil, a straight-edge and a sheet of paper the triangle is a geometrical concept: Infinitely thin with infinitely thin boundaries and infinitely sharp points, an object aethereal and without dimension. And yet they say its dimension is L^2 and that it possesses area.

My preferred "center" is the Gravitational Center or Gravicenter, which physicists call the Center of Gravity. It is a point of balance. For example, if you cut an arbitrary shape from a sheet of paper or metal and place it to lie on the point of a needle it would rest in perfect stable perpendicularity to the needle when you had arranged its Gravicenter to be exactly on the point of the needle. This sounds clear, but it is another theoretic ideal.

Notwithstanding all this, the concept of the triangle is central to quadrature, and quadrature is central to all human fabrications in industry, engineering and surveying, to civil and military planning, and of primary use to scientists and scholars in their researches. The integrated circuit has not reduced our need for quadrature: It has intensified it.

Any closed surface can be reduced, to some degree of precision, to a swarm of internal triangles. This is even true of curved surfaces like lakes, circles, ellipses and aeroplane wings. Of course, no straight-sided, flat form can represent a curve *exactly*, but precise approximation is usually good enough for practical people.

Even mathematical theorists sometimes find accurate methods of approximation useful, especially when dealing with advanced algebra involving equations which are impracticable or impossible to solve in an exact way.

If there are indeed an infinity of arrangements of triangles in any shape and those triangles each have an infinity of centers then it is a fair old guess that there are an infinity of methods of quadrature.

Mathematicians might be aware of a couple of dozen distinct methods of estimating areas, but a mere handful of these are useful in industry or the applied professions.

This is a simple book about simple methods that I can understand and that I hope you can find useful. I shall avoid all discussion of Delaunay-type optimisations of nets and nests of triangles: Of Finite Element Analysis: Of serious computer programming and anything else I lack the resources, mechanical and mental, to apply.

Once you have laid a straight datum line of known length, whether on paper, or with a surveyor's chain on the land, or using modern laser telemetry, you are in a position to form a triangle by sighting a line from each terminus to meet at some point hence where a colleague can set a standard (literally or figuratively) in the form of a calibrated pole.

Knowing one line and two angles you can then apply the Sine Rule to complete the other two lines. Army officers and other surveyors of the past did this personally, but now we would find aerial photogrammetry and computer data processing swifter, safer, more accurate and more economical.

Let us have an initial look at a blue-bordered blob whose area we wish to estimate:-

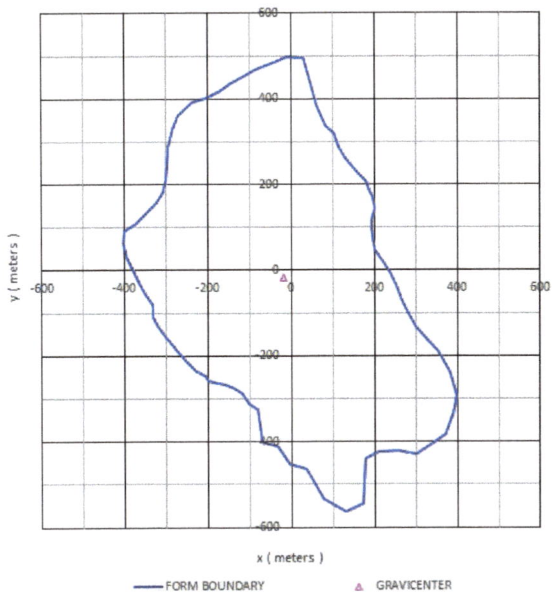

Figure 1.1
Our Blue-Bordered Blob

I shall call our blue-bordered blob Lake Ellesmere. You may think this a perfectly reasonable name, as good as any other. But to an Englishman, especially a certain type of Englishman, it is a horror.

You see, it is a pleonasm. *Mere* is the Old English word for "lake" or "pond" derived from the Latin *mare* for "sea". Therefore I am referring to "Lake Eels' Lake" where an eel is a type of edible fish, semiamphibious, quasi-marine, and once common in my country.

So in future I shall simply write of Ellesmere, and if I need to discuss the adjacent town I shall write "Ellesmere Town".

We may organise this and prepare to assess its area by the method of Diminishing Triangles in the way illustrated below.

Triangulation

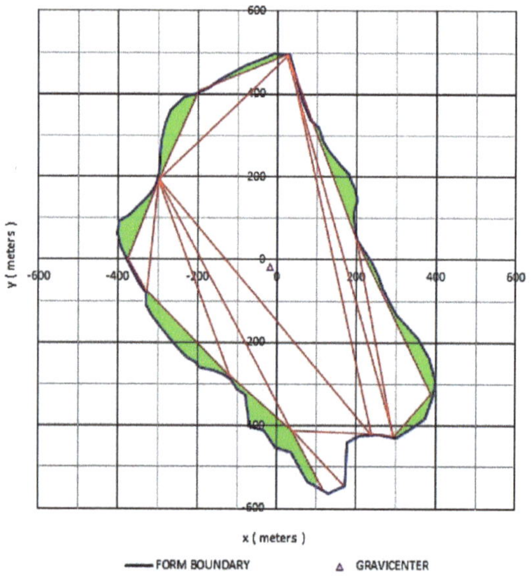

Figure 1.2
The Scheme of Diminishing Triangles

With a mere six "shore stations" and some eighteen sightlines you have accounted for roughly 95% of the lake's area, the nine green margins representing the area unaccounted for, but potentially resolvable with further stations and sightlines.

This is very similar to the methods that the officers and men of The Ordnance Survey of Great Britain would have used in 1867 when they first mapped Ellesmere scientifically.

They would judiciously have chosen shore stations to be more or less evenly distributed along the shore of the lake after carefully measuring the straightest segment with a sixty-six-foot surveyor's chain to set a fiducial baseline.

At each end they would have taken an angular measurement to a distant pole on the opposite shore using a theodolite, essentially a telescope on a precision turntable, calibrated to seconds of a degree. They would have done this at each end, and continued to sight further shore stations and their angles.

There is much to fault in *my* scheme.

Firstly, there should be a clear datum line replicable on land: Shore stations should be nearer to thirty in number to suppress statistical errors and more evenly distributed. Initial angular measurements, however, should be at roughly 120° ($2\pi/3$ radians) intervals about the lake to establish a roughly equilateral starting triangle which is incidentally likely to account for thirty or forty percent of a lake's surface.

Perhaps worst of all my lazy scheme includes many angles which are very acute or very obtuse inviting inaccurate angle readings and uncertainly in the length of lines: Phenomena known as "ill-conditioning". A skilled surveyor or competent geometer would avoid ill-conditioning at all costs, even if it meant traveling to almost inaccessible places.

After they had measured the angles of their conceptual triangles, diligent people would check that they totaled 180°. Victorian geographers would not have been happy with an error appreciably greater than a second of arc, which is 1/3600 of a degree or about 0.00000484813681109536 radians.

When you are coplanar with the object of your study, whether it is a pond or a galaxy, it is difficult to assess its area.

It is apparently easier if you have a perpendicular view, as from a bird, a drone or a satellite and can see the object conspectually.

I well remember a grim winter's day when reading for my first Master's degree I visited a cartographer at Glasgow University. I sought copies of aerial photographs, glossy black-and-white halide prints in those days, in order to commence my research into the planimetry of rivers. These pictures had been taken by the RAF during The Second World War to establish exactly what was where, because it was thought that we would have to fight the Nazis on our own land. After The War prints were made available to scientists and scholars.

My host attempted to dissuade me from using aerial photographs, recommending maps as more accurate. Certainly there was some very impressive technology crowding his laboratory, analog of course, and he explained how this was needed to un-distort photographs. My thoughts then and now were that one should recognise that to interpose technology between Nature and Reason was to insert Doctrine and assumption, and Doctrine brings Error. Men and women are the Heirs of Error and its authors as they bear and beget, nurture and guide. I kept my own counsel.

This kind if slightly patronising advisor gave me an excellent paper map of the Isle of Rhum as a parting gift, undoubtedly his work.

I hence forth digitised Ordnance Survey 1:25000 paper maps, purchased for me by the Science Research Council.

Counting the Squares

You possibly recall, with love or loathing, your school days.

Possibly, a geography mistress or your engineering drawing master set you the task of estimating an area by Counting Squares. Perhaps you were ordered to trace the outline of a lake or county from a map, or perhaps the profile of a cam or clutch-pad from a blueprint.

You may even have been persuaded to graticulate the outline of the subject with equally-spaced pencil lines a measured constant distance apart.

You counted the whole squares and multiplied by their pre-determined individual area and then tried to estimate the fractions of bordering squares intersected by the figure boundary, scaling these too by the whole-squares' area..

Figure 1.3 represents my attempt to do Counting Squares for Ellesmere. The image is graticulated with the small squares having a side length of 100 actual Ground Meters, and hence an area of 10000 square meters or one hectare for each square.

I have marked the whole lacustrine squares with red number-ones and attempted to guesstimate the areas of water in each boundary square. The tiny purple triangle is at the gravicenter, in this case very near to the assumed center at the origin, which I chose subjectively and is not necessarily the gravicenter for the quadrature purpose.

Meanwhile the feint green circle, centered on the subjective origin, marks the circular area equivalent to the computed area of the subject form, or rather in this case the fiducial area of the lake established by the Ordnance Survey in 1867AD.

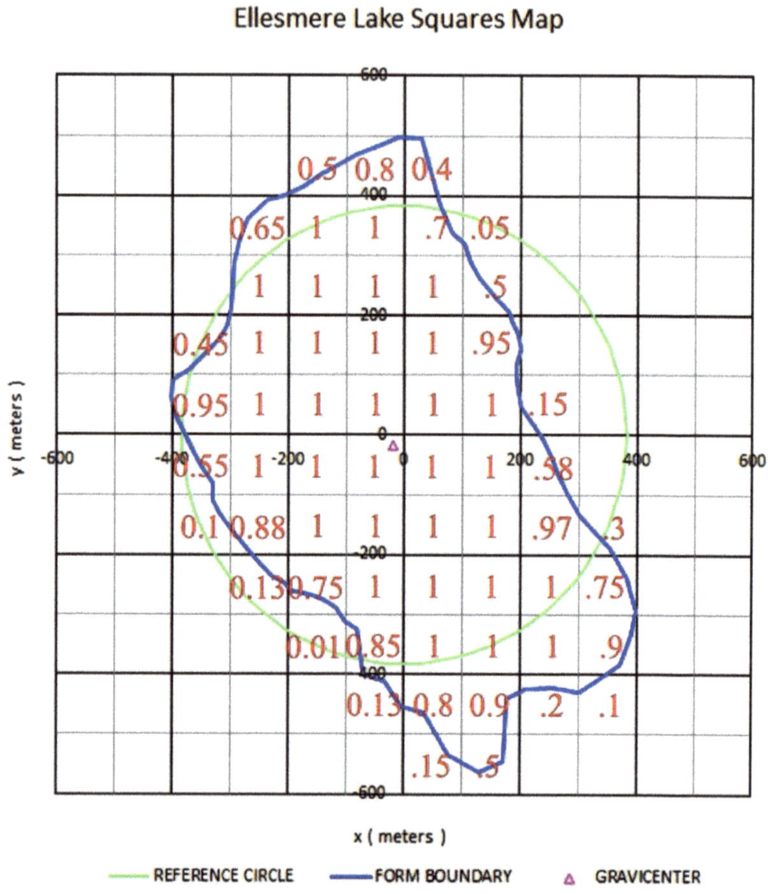

Figure 1.3
Counting the Squares of Ellesmere

For Ellesmere, this Fiducial Value is 460735 square meters and can be used to adjudge the accuracy of our own works and comparisons of the efficacy of different methods.

Table 1.1 presents the totals and comparisons for this Square Count of the quadrature of Ellesmere.

Note that as usual Percentage Specific Defect is our mathematical way of comparing the agreement of two figures. PSD or Percentage Specific Defect is defined by:-

$$PSD(x,y) = 100\left(\frac{x-y}{x}\right)$$

Equation 1.1

If PSD is positive then our estimate of area falls short of the fiducial area. In theory, any method that depends upon fitting straight-lines, however short, to a curving boundary must yield a positive PSD. In practice it is often otherwise.

By the way, there are more scientific ways of assessing the area of a transected square than judging by eye, but these ways are beyond the scope of this book.

If you have made judicious assessments of the areas of part-squares at the boundary then their average should be 0.5 units or 5000 square meters. In our case the relevant average is 5216.667 square meters.

For a theoretical x of 0.5 and a practical x of 0.5216667 the resulting PSD Error of Assessment is given by:-

$$PSD(0.5,0.5216667) = 100\left(\frac{0.5-0.5216667}{0.5}\right)$$
$$= -4.3334000000001$$

Equation 1.2

meaning that our judgment of graticule square part-areas is 4.33% too big.

At the end of the exercise our 31 whole squares account for 310000 square meters of the lake's surface and the thirty part-squares contribute a further 156500 so that our estimate of the total area is 466500 square meters.

To compare this with our nineteenth-century Ordnance Survey fiducial of 460735.011 m^2 we could re-employ the PSD equation as:-

$$PSD\left(A_{lakeOS}, A_{squarescount}\right) = 100\left(\frac{460735.011 - 466500}{460735.011}\right) = -1.25125915382194$$

Equation 1.3

So our Counting Squares area estimate is one and a quarter percent too big.

Surprisingly, we shall find that this method is twice as good as some of the computer estimates, mediated by "exact" mathematical equations, at least for Shropshire kettle lakes assessed at the hands of an experienced mathematician.

Ellesmere Lake							
A_{lakeOS}	460735	Meters^2					
Squares' Side Length (meters)	100	Meters2					
Squares' Common Area	10000	Meters^2					
A_{lakeOS}	460735	Meters^2					
$A_{squarescount}$	466500	Meters^2					
PSD(A_{lakeOS},$A_{squarescount}$)	-1.25126						
			Count of Complete Squares	Area of Complete Squares (m^2)	Serial	Fractional Area of Square (m^2)	Area of Square
			31	310000	0	0.4	4000
					1	0.7	7000
					2	0.05	500
					3	0.5	5000
					4	0.95	9500
					5	0.15	1500
					6	0.58	5800
					7	0.97	9700
					8	0.3	3000
					9	0.75	7500
					10	0.9	9000
					11	0.1	1000
					12	0.2	2000
					13	0.5	5000
					14	0.15	1500
					15	0.9	9000
					16	0.8	8000
					17	0.13	1300
					18	0.85	8500
					19	0.01	100
					20	0.75	7500
					21	0.13	1300
					22	0.88	8800
					23	0.1	1000
					24	0.55	5500
					25	0.95	9500
					26	0.45	4500
					27	0.65	6500
					28	0.5	5000
					29	0.8	8000
							156500 **Total**
							5216.667 **Mean**

Table 1.1
A Counting Squares Quadrature of Ellesmere

Pick's Theorem

A variant of counting the squares is to count the gridline intersections ("points") rather than the actual squares. The method assumes that your figure is a (irregular) polygon whose boundary vertices coincide with graticule intersections.

The Pick's Theorem equation is:-

$$A_{Pick} = I + \frac{B}{2} - 1$$

Equation 1.4

where I is the Number of Interior Grid Points within the Shape; and B is the Number of Grid Points at the Boundary.

Radial And Cartesian Treatments

Radial Mapping involves selecting a center of the figure whose area is to be estimated and measuring the angles and lengths of ray-lines relative to that center of the vertices or points at the figure's boundary.

This is similar to doing a 2π-radian centimetric scan within the shape rather like seamen with their scanning centimetric radar displays which they use to detect and distance unlit obstructions or enemies even in night and fog.

Experience with manual polar scanning of quasi-circular lakes shows that dividing the scanned area into more than 40 or 80 sectors offers the maximum accuracy in area estimation by-eye, whilst 80 sectors organised into a circular rose increase the dazzle in the surveyor's eyes without any appreciable benefit.

For example, Figure 1.4:-

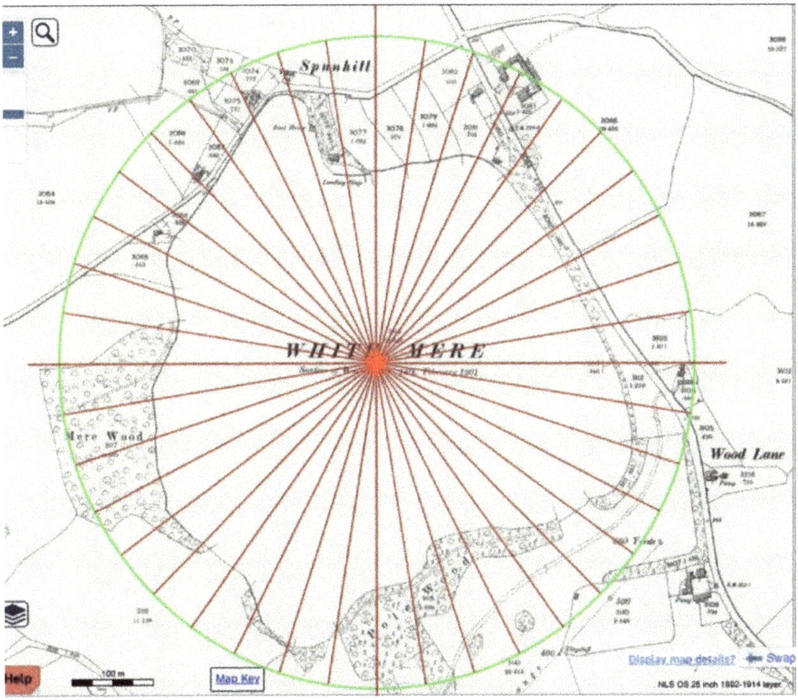

Figure 1.4
White Mere with a Superposed 40-interval Polar Rose Centered at an Arbitrary Origin

Figure 1.4 depicts another kettle lake near to Ellesmere. The map is taken from The National Library of Scotland Side-by-Side Internet Site and is a reproduction of part of a Victorian 25-inches-to-the-mile OS map.

The rose drafted by me has forty equal sectors and is operated by digitising the intersections of each successive red ray with the black lake shore to produce 41 polar co-ordinates. (Polar and Cartesian co-ordinates are mathematically interchangeable). For this example I just used my mouse cursor to read the Cartesian co-ordinates as I displayed the above image on my old but trusty PhotoDraw software.

You start (for consistency and convenience) with the Easternmost intersect and work sinistrally until you return to the same point. You take 41 readings to check the consistency of the first and the last.

With Cartesian treatments we do not bother with roses and polar co-ordinates, which demand the choice of some center of rotation, but instead process (x,y) Cartesian co-ordinate measurements directly.

The computed areas are mathematically identical and as always the choice of method, as in life itself, is a matter of ease, economy and accuracy.

Figure 1.5 represents a Cartesian method of boundary-point logging on Ellesmere using the online digitisation utility PlotDigitizer.

On Figure 1.5 you can see that I have invoked the online utility PlotDigitizer and loaded an image of the Ellesmere area as a mid-twentieth century OS 1:25000 map.

Note that PlotDigitizer is a Cartesian treatment and lists the successive black-circle co-ordinates on the left of the screen ready for commitment to a CSV file.

The eighty-interval rose was drafted by me and superposed upon the loaded image before mounting into PlotDigitizer. I intended the angular intervals dθ to be equal.

The thick arrowed blue line at the top is "exactly" 2000 meters long according to the Ordnance Survey graticulation and is used for cross-calibration purposes. Except in special circumstances, do not try to use the map's scale bar for calibrations: It is too short and you are too unsteady.

Figure 1.5
A Cartesian Digitisation about an 80-interval Rose using PlotDigitiser

PlotDigitizer internally calibrates when you move the utility's markers X1 and X2 to the western and eastern limits of the rose diameter. The 2000-meter blue line and the PlotDigitizer calibrations can be mathematically linked to render the recorded PlotDigitizer co-ordinates into meters.

I am an old man. The job really needs a young man or woman with a keen eye and a steady hand. Nevertheless I tried to click my mouse where the red rays and the black lake-shore meet, despite the dazzle.

The point is that radial Polar schemes of boundary digitisation are not necessarily superior to Cartesian ones, especially when human frailties intercede.

You can see that I have placed an initial black-circle at what I assumed to be the pole of the process, i.e. the nominal center of the lake. I have not been very accurate but I do not have to be as long as there are no islands or promontories that shadow a ray. As you return to the (easternmost) starting-point you should check that the figures match the second set on the list and click again to complete the circuit. If you want the gravicenter you can compute it from the data.

(A tiny island, Moscow Island, can just be seen in the South-West of the lake. It was created when waste from the local landscaping of Ellesmere House and its adjacent road was dumped in 1812AD. It pre-dates the OS survey by about thirty years. The OS estimated its acreage).

At the end you should have 82 digitised Cartesian points for an eighty-interval rose, plus the two PlotDigitizer calibration markers, and *take care to save your data* preferably in a .CSV, plus those two PlotDigitizer calibration markers.

Constrained by the equal sectors of the rose you can see that there is ill-conditioning in the form of a large boundary segment, ds, 4.5 to 9° East of North, and further ill-conditioning in the small bay 9 to 13.5° East of South.

Cartesian digitisations are robust in that they are not interfered with by re-entrant promontories or islands, but for an accurate estimate of the water surface you have of course to survey islands separately and deduct their areas.

You have not got to worry about the fact that all the ground straights curve due to The Curvature of The Earth's Oblate Spheroid when we are dealing with lines that are at most mere kilometers long: Rest assured, the sighting errors of yourself and the professional surveyors will more than overwhelm such corrections.

<u>Promontories and Islands</u>

As we inferred above, the adverse presence of promontories and islands will "cast a shadow" in Polar scanning schemes, just as ships' captains hope their centimetric microwave rays should.

Figure 1.6 illustrates this, for us, undesirable ray-shortening effect.

The green sectors are unaffected by hidden coastline behind promontories or islands in the blue-bounded lake whilst the red are affected, losing a considerable amount of shadowed area.

Cartesian-based methods such as Meister's Rule which do not need a center (Do not worry: You can calculate a gravicenter or another type of center of your choice from the returned data) are unaffected by promontories and islands, but you would have to calculate island areas separately.

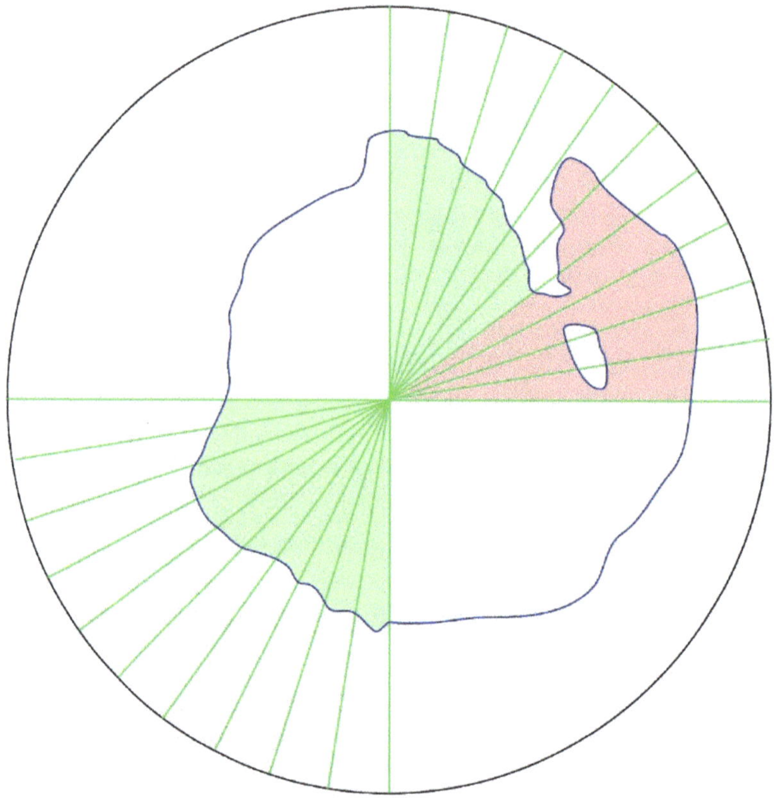

Figure 1.6
Vitiation of Polar Sectors by Islands and Promontories

The Lake of Menteith

There is a very hoary joke with which, predicably, I am about to bore you.

You see, there is only one lake in Scotland.

You may think: "How come? Scotland is a famous 'lakes and mountains' country. It has thirty thousand named bodies of subaerial natural fresh water."

Ah, it is like this. The Lake of Menteith is the only Scottish lake actually called a lake by Scots. The others are all "Loch" something such as Loch Ness and Loch Lochy, or "Lochan" if they are ponds. Further to confuse foreigners tidal inlets of the sea are also lochs like Loch Linnie and Loch Nevis, and of course you must never voice them as "lock", only as "lochch", with a voiceless velar fricative.

Conventional wisdom is that the Lake of Menteith is a kettle hole, a type of glacial lake formed by the melting of a residual ice block. We are hardly conventional geologists though.

Figure 1.7 is a Victorian Ordnance Survey map of the Lake of Menteith. I have traced the outline of the lake in black and colored it in light blue.

I have never visited the Lake of Menteith though I have driven past it many times.

I understand that, like many places in Scotland, it is a most beautiful and romantic location. It has the larger island of Inchmahome, which contains a ruined medieval Priory of the Black Canons that you can visit in Summer by regular motor ferry, or row your own boat whenever you please..

Let me quote the brilliant HV Morton as excerpted in the Wikipedia article "Inchmahome":-

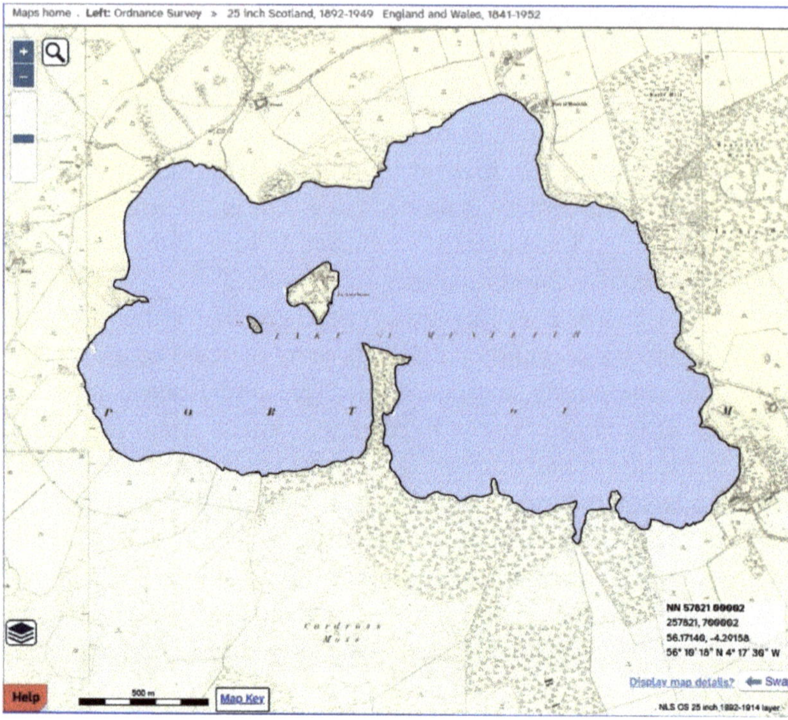

Figure 1.7
The Lake of Menteith

Far in the middle of the lake was a low greyness that rose and fell in queer shadows, as though the once holy isle of Inchmahome was built up out of lake water like a mirage.

This little island is one of Scotland's shrines; it was here that we can think of Mary Queen of Scots living, for once, happy hour[s] in that first garden of her simpleness.

On Inchmahome, she was hidden when the English were hammering at the gates of Stirling... Legend has labelled every yard of Inchmahome. There are the 'Queen's Bower', the 'Queen's Tree', and the 'Queen's Chamber'.

The mobile but star-crossed queen was only four when for weeks she was succored in the priory. There it is said, the Four Marys, Mary Beaton, Mary Fleming, Mary Livingston, and Mary Seton commenced her education in languages and the law. She would prove proficient in the former, but a poor judge of character.

The Lake of Menteith and its shores and islands are thickly wooded with thickets of gnarled old deciduous trees.

You can see that the Lake is a doubtful kettle, embracing as it does many promontories and three islands, albeit the smallest hardly affords standing room to a man.

The shore of the fabled lake comprises many quasi-circular embayments which seem to fall into two or three distinct diameters.

In Figure 1.8 I have taken the liberty of superposing red circles to model these hypothetical features. You can decern that there are five larger circles, five of middling size, four of which share the same diameters, and two little circles in the south-east.

To me these suggest swarms of meteoric ice fragments, perhaps from the Earth's brush with the tail of a passing comet, possibly in post-glacial pre-history.

I am not writing here of massive rogue asteroids or alien bodies of metal or rock that impact with an energy of gigatonnes and exterminate whole genera of creatures.

Such ice fragments would, for the larger 500-meter diameter astroblemes, be individually only 30-35 meters across, given their impact speed of around 25,000 miles per hour.

Sub-lacustrine contours and sampling of trace elements from lake sediments might throw light on this matter. A true kettle hole should have very abrupt depth fall-off near the shore, but have a flat bottom at a depth of only fifteen or so meters at the maximum.

The black lines represent suggested partitions for polar scans and the green crosses possible rose centers. The purple patches are areas of water in shadow from the mapped scheme. But any set of polar scans would only be practical from maps or aerial views of the Lake. And the islands remain a problem.

I disrecommend any polar schemes for this very irregular lake. Use Meister's Rule or something to estimate area from the shore.

If the Lake of Menteith has a shore length of 9700 meters and you aim for 1024 intervals then take GPS readings every 9700/1024 = 9.47265625 meters, say every nine and a half shore meters. (Tree shadow might be an issue).

If your readings are accurate to ±6 centimeters this does *not* mean that the interval should be either 0.06 or 0.12 meters: It means that your survey error *starts at* 6/950 meters or 0.6316%.

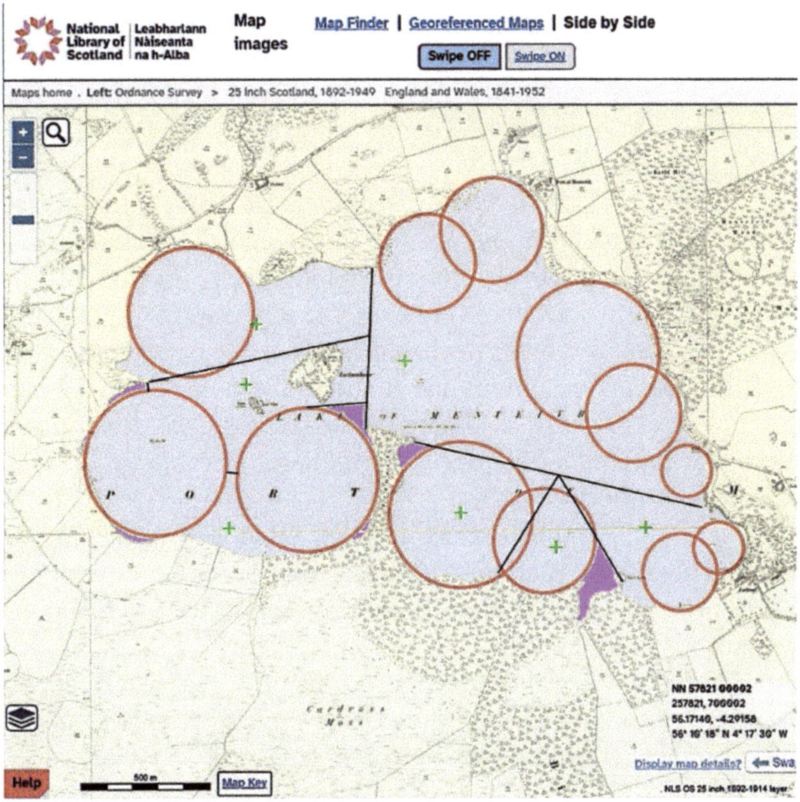

Figure 1.8
Embayment Circles and Sight Lines;
The Lake of Menteith

Planimeters

Planimeters are desk-based machines for measuring the outlines of shapes from maps or engineering drawings, etcetera.

Modern planimeters record the relative co-ordinates of clicked points on the document and compute the included area on demand. Victorian planimeters, for example of the AD1854 Amsler type, were sturdy brass instruments usually used to read relative areas of closed forms on the paper and could be calibrated against circles and squares of known physical area.

To use a planimeter make sure the supporting surface is perfectly flat, like a draftsman's drafting table, and adjustable for horizontality.

Planimeters are usually fitted with clear cross-haired cursors so that the operator may accurately pinpoint places to be clicked. Other designs may use needlepoint follower styluses.

When I digitised the courses of mapped rivers at the start of my PhD research I used my fingers to track a cursor over the lines. The cursor was of round glass and wired to a large flat table which had an electro-mechanical assembly underneath which grunted and ground disconcertingly as it followed the magnetic coil around the cursor. The digitised points were recorded to paper punched-tape, essentially tickertape, and I took these to another building which housed a mainframe computer with which I reproduced the river-course and analysed its conformation with programs I had written for the purpose.

For the modern user, life is much simpler.

Professional planimeters in 2026AD are about A5 in plan and no more than three centimeters in depth. They are easy to roll across any adequate flat desk and track either with a ball mechanically in contact with the paper or else a photoelectric transponder.

Figure 1.9 authored by Simona Fišnarová and taken from the Wikipedia article "Planimeter" shows a modern digital planimeter with an optical cursor at the top, and an Amsler with a needlepoint follower at the bottom. Decent modern digital planimeters are rated at ±0.1% accuracy. Expect to pay $1700 for a decent digital planimeter.

Photo: Simona Fišnarová

Figure 1.9
Digital, Hatchet and Amsler Planimeters

PART TWO
MATHEMATICAL FUNDAMENTALS

Polar and Cartesian Co-Ordinates Revisited

Figure 2.1 presents a Polar Sector and a peripheral Cartesian Increment set within the North-East Quadrant of a circle. For its Radius r, a perfect circle's area is πr^2, but this diagram is only so that we can demonstrate the Equivalence of Polar and Cartesian expressions and is quite general.

Please do not think that dx, dy and ds represent the vanishingly-small, theoretically infinitely-small, differentials of classical calculus. Our dx, dy and ds are actual measured increments of approximating straight lines at a boundary and each could be hundreds of meters long if you are recording the outline co-ordinates of a large county or whole country.

For this diagram, n, the Number of Polar Sectors is forty.

θ_6 is 60° or 1.0471975511966 radians, whilst θ_7 is 70° or 1.22173047639603 radians.

Accordingly, the equal Angular Increment, $d\theta$, is given by:-

$$d\theta = \theta_7 - \theta_6 = \frac{2\pi}{n} = 10° = 0.15707963267949\, radians$$

Equation 2.1

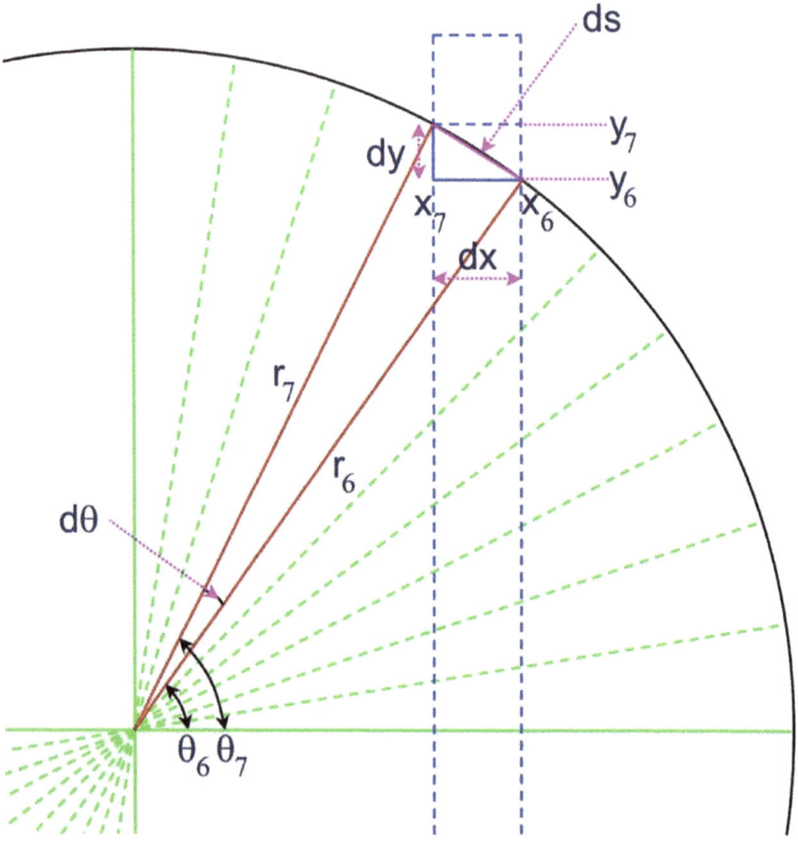

Figure 2.1
Polar and Cartesian Co-ordinates
Express One Another

The Cosine of dθ, Cosdθ, is 0.98768834059513 8 radians, whilst the Approximate Cosine of dθ, $Cosd\theta_{approx,}$ is given by:-

$$Cosd\theta_{approx} = 1 - \frac{d\theta^2}{2} = 0.987662994498638$$

Equation 2.2

Recollecting that the Percentage Specific Defect, PSD(x,y), is given by:-

$$PSD(x,y) = 100\left(\frac{x-y}{x}\right)$$

Equation 2.3

we compute, for an equal-interval 40-sector rose:-:-

$$PSD(\text{Cosd}\theta, Cosd\theta_{approx}) = 0.002566203878061$$

Equation 2.4

So it is clear that using a forty-interval rose with an approximate cosine of dθ is accurate to about one part in thirty-nine thousand. Unless your survey accuracy exceeds this you may as well effect the relevant computational saving.

<u>The Comparator Equation, J</u>

A common element of both the Polar system of quadrature and the Cartesian system is the Shape Boundary Chord ds of which a 40-interval rose would present forty examples, generally of different lengths.

ds is the purple line in Figure 2.1

Clearly, ds can be arrived at via Polar arithmetic or by Cartesian arithmetic, and should be numerically identical in either case. Allow that ds_{cart} is a Cartesian boundary element and ds_{polar} a Polar boundary chord. Then by the Theorem of Pythagoras:-

$$ds_{cart} = \sqrt{dx^2 + dy^2}$$

Equation 2.5

and by the Cosine Rule:-

$$ds_{polar} = \sqrt{r_6{}^2 + r_7{}^2 - 2.r_6.r_7.\cos(d\theta)}$$

Equation 2.6

Note also that:-

$$r_i = \sqrt{(x_i - x_{center})^2 + (y_i - y_{center})^2}$$

Equation 2.7

where i is a Sector Counter value.

Now ideally the Polar Comparator Equation, J, is:-

$$J = \frac{ds_{cart}}{ds_{polar}} = \frac{\sqrt{dx^2 + dy^2}}{\sqrt{r_6{}^2 + r_7{}^2 - 2.r_6.r_7.\cos d\theta}} = 1$$

Equation 2.8

For brevity allow that:

$$W = 2[x_{center}(x_{center} - x_6 - x_7) + y_{center}(y_{center} - y_6 - y_7)]$$

Equation 2.9a

$$U = \sqrt{(x_6 - x_{center})^2 + (y_6 - y_{center})^2} = r_6$$

Equation 2.9b

$$V = \sqrt{(x_7 - x_{center})^2 + (y_7 - y_{center})^2} = r_7$$

Equation 2.9c

U, V and W are functions of x_{center} and y_{center}.

The fact that J is unity demonstrates the essential redundancy of the shape centroids to quadrature and the superiority of Cartesian data for the quadrature application.

Where two processes are tantamount, the swiftest or cheapest is to be preferred. Usually the swiftest and cheapest method involves the fewer operations, especially expensive ones like roots or trigonometric functions.

So further:-

$$J = \frac{\sqrt{(x_6 - x_7)^2 + (y_6 - y_7)^2}}{\sqrt{(x_6{}^2 + y_6{}^2) + (x_7{}^2 + y_7{}^2) + W - 2.U.V.\cos(d\theta)}} = 1$$

Equation 2.10

Suppose that x_{center} and y_{center} are both zero, because the polar data has been pre-centered by one method or another (even closing your eyes and sticking a pin in the middle of your drawing). Then W is zero and U and V meaningless.

In real life, taking three-figure readings from your diagram using a PhotoDraw .mix file, the following is true:-

x_{center}	19.400000000000000
y_{center}	107.700000000000000
x_6	78.200000000000000
y_6	26.800000000000000
x_7	64.900000000000000
y_7	18.700000000000000
$d\theta$	0.157079632679490
$Cos(d\theta)$	0.987688340595138
r_6	100.011249367259000
r_7	99.956240425498200
dx	13.300000000000000
dy	8.100000000000000
ds_{polar}	15.689364265001600
ds_{cart}	15.572411502397400
ds_{cart}/ds_{polar}	0.992545729665732
$PSD(ds_{cart}, ds_{polar})$	-0.751025379634477
J	0.992545729665732
PSD(1,J)	0.745427033426782

Table 2.1
Summary of Comparator Equation Data

The computed 0.75% departure of J from unity is due to diagram measurement errors.

Data Preparation

When digitising your closed boundary data it is usually wise to link the software calibration to ground meters on the map or diagram before clicking the actual points.

Depending upon the digitisation program used this should enable you to record the co-ordinates directly as meters.

If you are digitising a map do not attempt to employ the Scale Bar. Calibrate against the graticules instead, and rely upon their separation. Aim to measure a line along the graticules that is rather longer than the longest dimension of the surface you are measuring. Where no graticules are shown you may, of course, have little choice but to rely upon the Scale Bar.

Recall Figure 1.5 and our discussion of the use of PhotoDraw with PlotDigitizer to record 80+ co-ordinates of the shore of Ellesmere.

Table 2.2 presents the recorded and computed calibration data for this exercise, whilst Table 2.3 is the same showing the orchestrating formulae:-

PhotoDraw Parameters					
	Scale Square Ground Value (meters)	2000			
			x	y	Length
	PhotoDraw Circle Diameter (mm)		200.02	200.02	200.02
	PhotoDraw Scale Bar Length (mm)		222.25		
	PhotoDraw Scale Ratio				8.998875
	PhotoDraw Circle Diameter in Ground Meters				1799.955
PlotDigitiser Parameters					
			x	y	
	PlotDigitizer Centroid		905.0478	899.9775	
	PlotDigitiser Scale Bar Point 1		0.633231	0.516129	0.081674
	PlotDigitiser Scale Bar Point 2		0.714906	0.516129	
	PlotDigitiser Scale Bar Point 3		0.633231	0.449565	0.079122
	PlotDigitiser Scale Bar Point 4		0.712353	0.448285	
	Average Scale Bar Length				0.080398
	PlotDigitizer Scale Ratio				12.4381
	Mean of Scale Ratios				10.71849
	Nominal Centroid (meters)		0	0	

Table 2.2
Digitisation Calibration Data for Ellesmere

	A	B	C	D	E	F
1	**PhotoDraw Parameters**					
2		Scale Square Ground Value (meters)	2000			
3				x	y	Length
4		PhotoDraw Circle Diameter (mm)		200.02	200.02	=(D4+E4)/2
5		PhotoDraw Scale Bar Length (mm)		222.25		
6		PhotoDraw Scale Ratio				=C2/D5
7		PhotoDraw Circle Diameter in Ground Meters				=C2*(F4/D5)
8						
9	**PlotDigitiser Parameters**					
10				x	y	
11		PlotDigitizer Centroid		905.047795774647	899.9775	
12		PlotDigitiser Scale Bar Point 1		0.63323124042879	0.516129032258064	=D13-D12
13		PlotDigitiser Scale Bar Point 2		0.714905532907015	0.516129032258064	
14		PlotDigitiser Scale Bar Point 3		0.63323124042879	0.449564803397417	=D15-D14
15		PlotDigitiser Scale Bar Point 4		0.712353272606001	0.448284658968174	
16		Average Scale Bar Length				=(F12+F14)/2
17		PlotDigitizer Scale Ratio				=1/F16
18						
19		Mean of Scale Ratios				=(F6+F17)/2
20		Nominal Centroid (meters)		=D11*E7	=E11*E7	

Table 2.3
Digitisation Calibration Data for Ellesmere
(EXCEL spreadsheet formulae)

The PhotoDraw circle diameter is the mean of the East-West and North-South diameters of the rose as it is superposed upon the PhotoDraw image which is intended to be 200 PhotoDraw millimeters.

The Scale Bar Length is the length of the Blue Line which in turn is 2000 ground meters so that the PhotoDraw Scale Ratio is =C2/D5 = 8.998875.

On the PlotDigitizer hand, cells F12 through F16 compute the respective PlotDigitizer Scale Ratio average as =(F12+F14)/2 = 0.080378.

The actual PlotDigitizer Scale Ratio is the reciprocal of this, 12.4381, leading to an inter-utility mean of 10.71849.

But this is essentially a post-facto report of parameters, because the rigmarole of X1, X2 and Y1, Y2 mappings, taken with the 2000-meter map scale definition, ensured that PlotDigitizer readings would report in ground meters, as illustrated by the part-tabulation of Table 2.4.

Serial	RAW ex PlotDigitizer		RAW without Centroid Header Line		PlotDigitizer Points Without Repeats		Serial	PlotDigitiser Consolidation with Closure		Serial	Digitised Boundary Points in Ground Meters	
	x	y	x	y	x	y		x	y		x	y
0	905.0478	899.9775	1143.352	894.9356	1143.352	894.9356	0	1143.352	894.9356	0	1143.352	894.9356
1	1143.352	894.9356	1128.141	917.6241	1128.141	917.6241	1	1128.141	917.6241	1	1128.141	917.6241
2	1128.141	917.6241	1120.535	927.7079	1120.535	927.7079	2	1120.535	927.7079	2	1120.535	927.7079
3	1120.535	927.7079	1107.86	947.8755	1107.86	947.8755	3	1107.86	947.8755	3	1107.86	947.8755
4	1107.86	947.8755	1102.789	975.6059	1102.789	975.6059	4	1102.789	975.6059	4	1102.789	975.6059
5	1102.789	975.6059	1100.254	995.7734	1100.254	995.7734	5	1100.254	995.7734	5	1100.254	995.7734
6	1100.254	995.7734	1100.254	1018.462	1100.254	1018.462	6	1100.254	1018.462	6	1100.254	1018.462
7	1100.254	1018.462	1107.86	1043.671	1107.86	1043.671	7	1107.86	1043.671	7	1107.86	1043.671
8	1107.86	1043.671	1102.789	1071.402	1102.789	1071.402	8	1102.789	1071.402	8	1102.789	1071.402
9	1102.789	1071.402	1092.649	1091.569	1092.649	1091.569	9	1092.649	1091.569	9	1092.649	1091.569
10	1092.649	1091.569	1085.043	1106.695	1085.043	1106.695	10	1085.043	1106.695	10	1085.043	1106.695
11	1085.043	1106.695	1064.762	1126.863	1064.762	1126.863	11	1064.762	1126.863	11	1064.762	1126.863
12	1064.762	1126.863	1052.086	1144.509	1052.086	1144.509	12	1052.086	1144.509	12	1052.086	1144.509

Table 2.4
Adjustment of the first 12 of 81
Recorded Ellesmere Co-Ordinates
As Recorded in Ground Meters in a .CSV File

The zeroth point in the RAW columns is the assumed lake center which I clicked redundantly for the benefit of polar treatments. Subsequently I dispensed with that (a centroid could later be computed from boundary data, if required).

I then copied the first co-ordinate to the last to enforce shape closure, and after weeding-out the repetitions of this hand-and-eye digitisation, moved forward to produce a refined list of 80 cartesian boundary co-ordinates in the x and y columns of "Digitised Boundary Points in Ground Meters".

It is this fifth pair of columns which can then be committed to a .CSV file and fed to whichever program we will apply to areal and boundary analysis.

As you can see the Cartesian ground co-ordinates hover around one kilometer which is the correct order for the Shropshire lake of Ellesmere.

Some Equations for Polar Area

It is useful first of all to decide the precision of our area and perimeter estimations in terms of the number of boundary points we shall record, bearing in mind that these should roughly be equally-spaced along the perimeter.

Sensible choices depend upon the elongation of the shape and the complexity of its perimeter, but choices of the Number of Points, n, will vary from a minimum of 40 to maybe 1024 or more depending upon the degree of accuracy tenable and the resources you can deploy.

In Polar terms, the whole sweep of your scan will be 2π radians, and accordingly:-

$$d\theta = \frac{2\pi}{n}$$

Equation 2.11

where dθ is the Angular Increment of Sampling, i.e. clicking on the boundary. Hence Σdθ is 2π.

Where practicable, the dθ's should be equal so that they may be tabulated using:-

$$\theta_i = i.d\theta$$

Equation 2.12

Continuing our discussion in Polar terms, the Local Radius of (Boundary) Intersection, r_i, is given by:-

$$r_i = \sqrt{x_i + y_i}$$

Equation 2.13

For centered co-ordinates (x_i, y_i).

Thus for the Sectoral Total Area Estimate, A_{sector}:-

$$A_{sector} = \sum_{k=0}^{n-1} dA_k$$
$$= \sum_{i=0}^{n-1} \frac{1}{2} r_i r_{i+1}.\sin(d\theta_i) = \frac{1}{2} \sum_{i=0}^{n-1} r_i r_{i+1}.\sin(d\theta_i)$$

Equation 2.14

or for equal dθ's:-

$$A_{sector} = \frac{1}{2}.\sin(d\theta) \sum_{i=0}^{n-1} r_i r_{i+1}$$

Equation 2.15

dA_k is the Sectoral Contribution to the Total Area of the shape.

Meister's Rule

Meister's Rule is a direct and efficient method of estimating area by reference to *the list of x and y co-ordinates only.*

Meister's Rule makes no assumption about the center of the figure though a center may be calculated separately if required.

In the literature, you will often find that Meister's Rule is called "The Shoelace Formula", the "Surveyor's Formula" or something. The relevant Wikipedia article is entitled "The Shoelace Formula".

The formula for the Estimate of Total Figure Area by means of Meister's Rule, $A_{Meister}$, is:-

$$A_{Meister} = \sum_{j=1}^{n-1} y_j . \left(x_{j-1} - x_{j+1}\right)$$

Equation 2.16

Simpson's Rule

Simpson's Rule is an incredibly useful and practical method of numerical quadrature otherwise called numerical integration. It is simple, flexible, and if judiciously applied, highly accurate.

It is a member of a family of similar mathematical processes called Newton-Cotes formulae, and is a "closed" formula that assumes that your data set stands complete.

Simpson's Rule effectively assumes that area-estimation data-points are connected by curving segments of a quadratic equation. This is strictly false for any mathematical structure other than actual second-degree algebraic equations, yet ironically quadratic curves are excellent fits to any arbitrary boundary as long as your selected interval is short enough. If you divide the boundary of any ideal closed surface into 1024 intervals (i.e. 2^{10} increments) then you will be almost certain to achieve twelve or fifteen digits of denary accuracy.

Obviously you cannot do such fine work by manually-reading a polar rose and clicking a mouse, but computerised equipment can process such data.

Simpson's Rule is named for Thomas Simpson (1710-1761), a self-taught English mathematician. To what extent Simpson was original is a matter for debate as both Kepler and Cavalieri were aware of similar area-estimation rules one hundred years previously.

In 1615 the great German astronomer Johannes Kepler published *Keplersche Fassregal*, literally "Kepler's Barrel Rule" which was used by coopers and their customers to estimate the capacity of wooden stave barrels ubiquitously used during the seventeenth and eighteenth centuries for the transport of bulk liquids.

This most basic form of Simpson's Rule is given by:-

$$\int_a^b f(x) \approx \frac{b-a}{3}\left[f(a) + 4f\left(\frac{b-a}{2}\right) + f(b)\right]$$

Equation 2.17

which fits a parabola to two points a and b which are b-a units apart. The algebra on the left hand side (LHS) of the approximately-equal sign is called the General Figure of Integration and essentially declares that the right-hand side (

RHS) is the area enclosed under the line y = f(x), a surface called by mathematicians "the integral". The General Figure of Integration is tantamount to area.

An application of Simpson's Rule must always involve an even number of equal intervals.

Therefore the next most sophisticated (and hopefully accurate) estimator of area is:-

$$\int_a^b f(x) \approx \frac{h}{3}[f(x_0) + 4f(x_1) + 2f(x_2) + 4f(x_3) + f(x_4)]$$

Equation 2.18

where h is the Common Interval, and $f(x_i)$ are successive y-values.

This is an Extended Simpson's Rule, extended in the sense that we have articulated a second series of terms to the back of the series and "overlapped" the last term of Equation 2.17 to give the multiplier 1+1=2.

Between i = 1 and i = n-1 the term multipliers alternate 4,3,4,2,4,2,...,4 so that the inner series always starts and ends with a 4.

Therefore, to generalise we may write:-

$$\int_a^b f(x) \approx \frac{h}{3}\left[f(x_0) + 4\sum_{i=1}^{n-2} f(x_{2i-1}) + 4\sum_{i=1}^{n/2-1} f(x_{2i}) + f(x_n)\right]$$

Equation 2.19

The structure of Equation 2.19 is not always computationally convenient or economical.

Thirty or forty years ago I noticed that the series 4,2,4,2,... could be expressed by this integer alternator:-

$$ALT(i) = 3 + (-1)^{i+1}$$

Equation 2.20

Accordingly, Equation 2.19 can be re-cast as:-

$$\int_a^b f(x) \approx \frac{h}{3}\left[f(x_0) + \cdot \sum_{i=1}^{n-1} (ALT(i).f(x_i)) + f(x_n)\right]$$

Equation 2.21

which is convenient for any even number of intervals.

For an eight-interval Simpson Integration applied to an ellipse, Table 2.5 presents illustrative output.

As we observed at the outset of this discussion of Simpson's Rule it is only exact for second-degree polynomial curves, that is to say parabolas.

And yet by one of those ironies of Life and Science, the parabola is the best approximant of any curve as long as the chord is sufficiently short.

The parabola is an artificial construct and matches no curve but itself. Indeed it is algebraic and it is therefore impossible to fit to a transcendental function. Both the perimeter of a circle and the perimeter of an ellipse are functions of a transcendental number, The Ludolphine Constant, π (pi).

And yet, as we shall see, a parabola is an almost matchless fit for a piecemeal curve.

I say an "almost matchless" fit because in mathematics and science there is, as in life, always room for improvement.

Simpson Integration			
Serial	ALT_{kk}	g_{kk}	Simpson Term
0	1	2.618034	2.618034
1	4	1.447214	5.788854
2	2	1	2
3	4	1.447214	5.788854
4	2	2.618034	5.236068
5	4	1.447214	5.788854
6	2	1	2
7	4	1.447214	5.788854
8	1	2.618034	2.618034

Table 2.5
An Example of an Eight-Interval Simpson Integration in a Spreadsheet

Indeed, I hope you improve upon my methods and any other proposal made.

Testing Our Methods

It is most important that we test our methods, and especially their instantiations as spreadsheets or computer programs before we attempt to apply them to the environment.

It is life-critical that we test our methods by three independent avenues before we apply them to medical or military procedures, or to engineering design.

Any tool than assists testing, whether stochastic (statistical) or deterministic should be considered.

In life-critical projects the lives of men and women are on the line. No expense should be spared, and if the tool is foreign or even enemy, we should put our pride in our pocket and exploit it, after due diligence: We all remember the story, possibly apocryphal, of the British aircraft carrier blueprint that was stolen by the Japanese who copied it religiously, only to watch their mighty vessel capsize as it left the slipway.

The Percentage Specific Defect

The Percentage Specific Defect is a measure of accuracy.

I can hardly improve upon the Google AI Overview of the distinction between Accuracy and Precision, so I shall quote it verbatim:-

AI Overview

Accuracy refers to how close a measurement is to the true or accepted value, while precision refers to how close multiple, repeated measurements are to each other. High accuracy means low systematic error, whereas high precision indicates low random variability and high reproducibility.

Key Differences and Characteristics:

- ***Definition:*** *Accuracy is about correctness (closeness to the target), while precision is about consistency (closeness of measurements to each other).*
- ***Target Analogy:*** *Imagine a dartboard:*
    - ***Accurate:*** *Darts are clustered near the bullseye.*
    - ***Precise:*** *Darts are tightly clustered together, even if they are far from the bullseye.*
- ***Independence:*** *It is possible to have high precision but low accuracy (tight grouping far from the target) or*

low precision but high accuracy (spread out but averaging to the target).

- ***Context:*** *In scientific measurement, precision is dependent on the resolution of the instrument, while accuracy is dependent on proper calibration.*

Summary Table

Feature	*Accuracy*	*Precision*
Focus	*Correctness (True Value)*	*Consistency (Reproducibility)*
Question	*How close is it to the target?*	*How close are the measurements to each other?*
Error Type	*Minimizes Systematic Error*	*Minimizes Random Error/Variance*
Example	*Weighing 1kg item and getting 1.00kg*	*Weighing 1kg item 3 times and getting 0.90kg, 0.90kg, 0.90kg*

These facts are illustrated by the common analogy of arrow or bullet marks upon a target board, as in the diagram provided by https://sciencenotes.org/what-is-the-difference-between-accuracy-and-precision/ reproduced as Figure 2.2:-

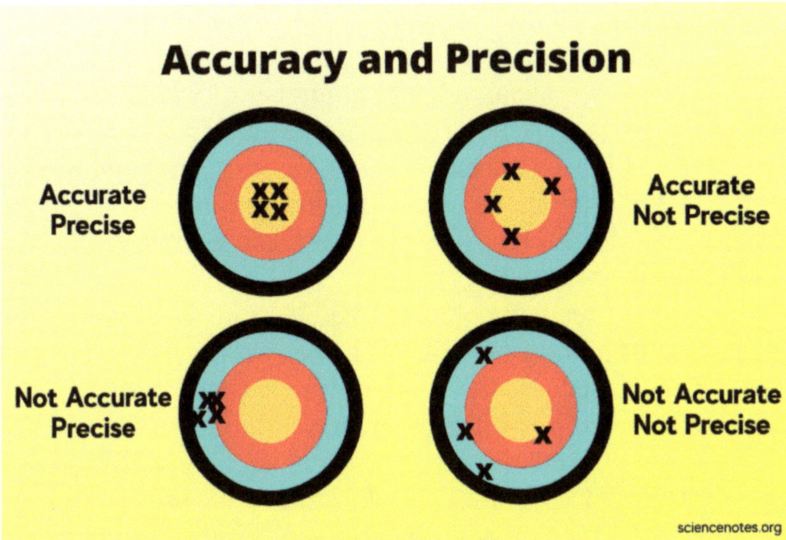

Figure 2.2
Accuracy and Precision

The Percentage Specific Defect is the simplest mathematical tool for comparing the relative agreement of two numbers.

It is defined by the function:-

$$PSD(x,y) = 100\left(\frac{x-y}{x}\right)$$

Equation 2.22

Percentage Specific Defect should be as small as possible, if x and y are thought to be equivalent.

If PSD is negative then y is too great an estimate of x: If PSD is positive, y is too small.

Fiducial Values and the Reference Circle

A fiducial value is a number in which we place our faith. Faith is an Act of Trust, but like all trust it is subject to trial and experience. Repeatability is key to the adoption of a fiducial value. The agreement with a fiducial number is about accuracy, not about precision, because it can be an error which is consistently replicated.

For example, if the Nassau Experiment is repeated a million times all over the world and the same outcomes supersede in the same order on every occasion our faith in the procedure's integrity grows.

On the other hand, if men employed by a given organisation visit the Moon on two or three occasions and no-one else does so for sixty years, then our faith in the reality of the adventures takes a knock, just as it does with alleged miracles of the Church, even when our Faith in Christ is total.

Of a different order is our faith in the integrity of British officers and their technical methods. In the course of the nineteenth century the topographic features of India and Britain itself were surveyed by the Ordnance Survey, and re-surveyed when enemy occupation was again anticipated. This was done of course to assist tactical planning and resistance.

Of interest to us was the fact that the area and sometimes the depth profile of lakes was measured from field triangulations or soundings and the results published on maps after the threats of Napoleon or Hitler had receded.

Of a third order of fidelity is the number value that is the calculated outcome of an ancient and very well tested numerical procedure based upon geometrical theory, such as the Ludolphine Constant based ultimately upon the writings and drawings of Archimedes, a Sicilian Greek who strove to defend his home city of Syracuse from the Roman Navy.

Theoretical derivations are ideal.

Examples include the Celerity of Light and the Universal Gravitational Constant these two latter being *physical* rather than *mathematical* parameters and thus subject to continued test and experience, merely provisional.

I suppose you concur with me in agreeing that the perimeter of a circle is $2\pi r$ and its area πr^2 where r is the figure's radius, and that force diminishes with the inverse square of separation, but Archimedes and Newton were both *physicists* and they and their arts fallible: Einstein corrected them a little and he too shall be corrected in time. Inaccuracy or fallibility do not necessarily render a value useless. We are mortal men and women and we err of necessity, but we Work and we Provide knowing that we shall be bettered.

A reference circle is a fiducial graphic shape calculated to have the same area as that of whatever it is that you intend to assess. Plot it on your map or blueprint for another handy if inexact check on the credibility of your procedures.

The reference circle is defined by:-

$$r = \sqrt{\frac{A_{fido}}{\pi}}$$

Equation 2.23

where A_{fido} is the Fiducial Area, r is the Fiducial Radius and π is the Ludolphine Constant whose value is near to 3.14159265358979. The Ludolphine Constant is the theoretic ratio of a circle's circumference (perimeter) to its diameter and was first estimated in the Occident by Archimedes and other Hellenistic mathematicians around the year 300BC. The concept was known to other cultures in the Orient. The approximation 22/7 is frequently used.

The Cartesian co-ordinates of a circles' perimeter are given by:-

$$x_c = r.\cos(\theta)$$
Equation 2.24a

$$y_c = r.\sin(\theta)$$
Equation 2.24b

where (x_c,y_c) is a specific Fiducial Circle Circumferential Co-ordinate and θ is the Lapse Angle where in our convention θ = 0 radians at the Easternmost point of the circle.

The Fiducial Ellipse

The Ellipse is a deformed circle and its mathematical properties are very well established.

It is an ideal first test of any program of quadrature, ideal in both senses of the word, because its properties, whilst Transcendental, can be computed to arbitrary accuracy. Using my computer system, fifteen figures of accuracy is normal. Nothing measured in the field, or the microscope slide, can approach this level of exactitude.

The Area of the Ellipse, $A_{ellipse}$, is given by:-

$$A_{ellipse} = \pi ab$$
Equation 2.25

a and b are respectively the lengths of the (x) Major Axis and the (y) Minor Axis of the Ellipse, analogous to radius of the circle where a=b=r.

Simple, isn't it?

Regrettably, the Perimeter of an ellipse is not so straight forward.

We must, to "know" the value of the ellipse perimeter, compute a Complete Elliptic Integral of the Second Kind.

Firstly though we should compute a definitive property of our ellipse called the Eccentricity, e, defined by:-

$$e = \sqrt{1 - \frac{b^2}{a^2}}$$

Equation 2.26

Essentially, e is a metric of oblation.

Having e we can move forward.

The Complete Elliptic Integral of the Second Kind, E(e), is given by:-

$$E(e) = \int_0^{\pi/2} \sqrt{1 + e^2 . \sin(\theta)^2} \, . d\theta$$

Equation 2.27

There is no analytic solution to this integral, even though it is a definite integral, but a numerical value for a given e can be calculated numerically. Why don't you try Simpson's Rule?!

Now we can estimate the Ellipse Circumference, $C_{ellipse}$, using:-

$$C_{ellipse} = 4a. E(e)$$

Equation 2.28

You may see in some literature Ramanujan's First Approximation (1914):-

$$C_{approx} \approx \pi(a+b)\left(1+\frac{3h}{10+\sqrt{4-3h}}\right)$$

Equation 2.29

where:-

$$h = \frac{(a-b)^2}{(a+b)^2}$$

Equation 2.30

or alternatively:-

$$C_{appox} \approx \pi\left[3(a+b)-\sqrt{(3a+b)(a+3b)}\right]$$

Equation 2.31

Depending upon your application, you may care to investigate the accuracy and economy of these approximations.

I have access to powerful computer resources and henceforward will work with the "exact" solution involving the elliptic integral.

In polar terms the Radius of an Ellipse, r(θ), at Lapse Angle θ is:-

$$r(\theta) = \frac{ab}{\sqrt{(b.\cos\theta)^2 + (a.\sin\theta)^2}} = \frac{b}{\sqrt{1-(e.\cos\theta)^2}}$$

Equation 2.32

(The Lapse Angle is reconned sinistrally from θ = 0 at the Easternmost point).

What remains now is to select a and b values for our testing ellipse.

It is sensible that one of a or b is unity.

I propose that the shorter semi-axis b is unity and the longer a is The Major Ratio of Phidias, Φ, which has the equation:-

$$\Phi = \frac{1+\sqrt{5}}{2} = 1.61803398874989$$

Equation 2.33

For this ellipse fiducial shape the EXCEL-computed Eccentricity, e, is 0.786151377757423; the Elliptic Integral E(e) is 1.288568294506470; the Area $A_{ellipse}$ is 5.083203692315260; and the Circumference, $C_{ellipse}$, of the figure is 8.339789189347810. All of these are in arbitrary units relative to b = 1.

Figure 2.3 illustrates this ellipse together with its reference circle. As you may adjudge the blue circle and and the red ellipse are of at least approximately equal areas.

By analogy with the circle, a rough-and-ready estimate of the Area of the Ellipse $A_{\pi r2}$ is given by:-

$$A_{\pi r2} = \pi . r_{\mu}{}^{2}$$

Equation 2.34

where r_μ is the Mean Ellipse Radius defined by:-

$$r_{\mu} = \frac{\sum_{i=0}^{n+1} r_i}{n+1}$$

Equation 2.35

where n+1 is the Number of Radii.

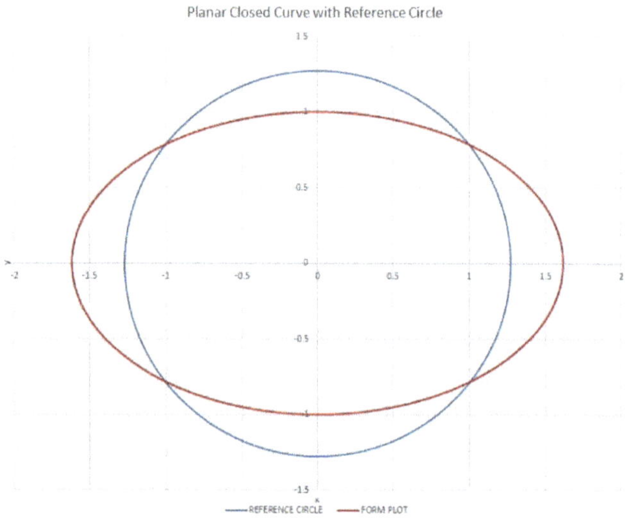

Figure 2.3
The Fiducial Ellipse and its Reference Circle

Equation 2.34 is unsatisfactory where n>16.

Table 2.6 shows the key input and outcome data comparing methods of simple quadrature for the case of 32 equally-spaced radial intervals of the 1:Φ ellipse.

You can see that already there is less than a one percent defect relative to the fiducial equation for Summation of Sectors, but that the Simpson Integration is better than that by a good three orders of magnitude. The Summation of Sectors is equivalent to Meister's Rule, but Estimation from Mean Radius is at about 1.12 PSD considerably less accurate than either.

Table 2.7 shows the step-wise outputs for these thirty-two sectors of the ellipse.

Number of Radial Intervals	32
Number of Perimeter Points	33
Range (2π)	6.283185307179590
dθ	0.196349540849362
Sine of dθ	0.195090322016128
Φ	1.618033988749890
a	1.618033988749890
b	1
ecc	0.786151377757423
E_{ecc} CEI 2nd Integral	1.288568294506470
θ_0	0.000000000000000
θ_n	6.283185307179590
Simpson Interval, h	0.196349540849362
Mean Radius	1.264886416167880
$\pi^*\mu_r^2$	5.026352354265920
$A_{Meister}$	5.038654032622900
$A_{ellipse}$	5.083203692315260
$A_{sectors}$	5.038654032622900
$A_{Simpson}$	5.083171008682530
PSD($A_{ellipse}$,$A_{sectors}$)	0.876409099240181
PSD($A_{ellipse}$,$A_{Simpson}$)	0.000642973107248
PSD($A_{ellipse}$,$A_{Meister}$)	0.876409099240181
PSD($A_{ellipse}$,$A_{\pi r^2}$)	1.118415501139350
$C_{ellipse}$	8.339789189347810
C_{approx}	8.316155283745960
PSD($C_{ellipse}$,C_{approx})	0.283387326289194
Radius of Reference Circle	1.272019649514070
Area of Reference Circle	5.083203692315260

Table 2.6
Comparative Table of the Inputs and Outcomes of a 32-Interval Quadrature of a 1:Φ Ellipse

Serial	θ	dθ	r	$r_k * r_{k+1}$	dA_k	x	y	Meister Term	dx	dy	ds
0	0	0.19635	1.618034	2.540963	0.247858652	1.618034	0		-0.07781	0.30637	0.316096
1	0.19635	0.19635	1.570402	2.284659	0.222857385	1.540227	0.30637	0.083931	-0.19614	0.250367	0.318051
2	0.392699	0.19635	1.454824	1.922368	0.187517742	1.344083	0.556737	0.245824	-0.2454	0.177379	0.302794
3	0.589049	0.19635	1.321375	1.589617	0.155059399	1.098683	0.734117	0.362236	-0.24803	0.116534	0.274044
4	0.785398	0.19635	1.203002	1.337297	0.130446889	0.850651	0.850651	0.409242	-0.23306	0.073639	0.244417
5	0.981748	0.19635	1.111634	1.165635	0.113702026	0.617591	0.92429	0.415355	-0.21632	0.04447	0.220841
6	1.178097	0.19635	1.048578	1.061133	0.103508355	0.401273	0.96876	0.407038	-0.20385	0.023768	0.205228
7	1.374447	0.19635	1.011973	1.011973	0.09871306	0.197426	0.992528	0.398275	-0.19743	0.007472	0.197567
8	1.570796	0.19635	1	1.011973	0.09871306	6.13E-17	1	0.394852	-0.19743	-0.00747	0.197567
9	1.767146	0.19635	1.011973	1.061133	0.103508355	-0.19743	0.992528	0.398275	-0.20385	-0.02377	0.205228
10	1.963495	0.19635	1.048578	1.165635	0.113702026	-0.40127	0.96876	0.407038	-0.21632	-0.04447	0.220841
11	2.159845	0.19635	1.111634	1.337297	0.130446889	-0.61759	0.92429	0.415355	-0.23306	-0.07364	0.244417
12	2.356194	0.19635	1.203002	1.589617	0.155059399	-0.85065	0.850651	0.409242	-0.24803	-0.11653	0.274044
13	2.552544	0.19635	1.321375	1.922368	0.187517742	-1.09868	0.734117	0.362236	-0.2454	-0.17738	0.302794
14	2.748894	0.19635	1.454824	2.284659	0.222857385	-1.34408	0.556737	0.245824	-0.19614	-0.25037	0.318051
15	2.945243	0.19635	1.570402	2.540963	0.247858652	-1.54023	0.30637	0.083931	-0.07781	-0.30637	0.316096
16	3.141593	0.19635	1.618034	2.540963	0.247858652	-1.61803	1.98E-16	8.8E-32	0.077807	-0.30637	0.316096
17	3.337942	0.19635	1.570402	2.284659	0.222857385	-1.54023	-0.30637	0.083931	0.196144	-0.25037	0.318051
18	3.534292	0.19635	1.454824	1.922368	0.187517742	-1.34408	-0.55674	0.245824	0.245399	-0.17738	0.302794
19	3.730641	0.19635	1.321375	1.589617	0.155059399	-1.09868	-0.73412	0.362236	0.248032	-0.11653	0.274044
20	3.926991	0.19635	1.203002	1.337297	0.130446889	-0.85065	-0.85065	0.409242	0.23306	-0.07364	0.244417
21	4.12334	0.19635	1.111634	1.165635	0.113702026	-0.61759	-0.92429	0.415355	0.216317	-0.04447	0.220841
22	4.31969	0.19635	1.048578	1.061133	0.103508355	-0.40127	-0.96876	0.407038	0.203847	-0.02377	0.205228
23	4.516039	0.19635	1.011973	1.011973	0.09871306	-0.19743	-0.99253	0.398275	0.197426	-0.00747	0.197567
24	4.712389	0.19635	1	1.011973	0.09871306	-1.8E-16	-1	0.394852	0.197426	0.007472	0.197567
25	4.908739	0.19635	1.011973	1.061133	0.103508355	0.197426	-0.99253	0.398275	0.203847	0.023768	0.205228
26	5.105088	0.19635	1.048578	1.165635	0.113702026	0.401273	-0.96876	0.407038	0.216317	0.04447	0.220841
27	5.301438	0.19635	1.111634	1.337297	0.130446889	0.617591	-0.92429	0.415355	0.23306	0.073639	0.244417
28	5.497787	0.19635	1.203002	1.589617	0.155059399	0.850651	-0.85065	0.409242	0.248032	0.116534	0.274044
29	5.694137	0.19635	1.321375	1.922368	0.187517742	1.098683	-0.73412	0.362236	0.245399	0.177379	0.302794
30	5.890486	0.19635	1.454824	2.284659	0.222857385	1.344083	-0.55674	0.245824	0.196144	0.250367	0.318051
31	6.086836	0.19635	1.570402	2.540963	0.247858652	1.540227	-0.30637	0.083931	0.077807	0.30637	0.316096
32	6.283185		1.618034			1.618034	-4E-16				

Simpson Integration

Serial	ALT_{kk}	g_{kk}	Simpson Term
0	1	2.618034	2.618034
1	4	2.466161	9.864644
2	2	2.116514	4.233028
3	4	1.746032	6.984126
4	2	1.447214	2.894427
5	4	1.235729	4.942917
6	2	1.099516	2.199032
7	4	1.024089	4.096357
8	2	1	2
9	4	1.024089	4.096357
10	2	1.099516	2.199032
11	4	1.235729	4.942917
12	2	1.447214	2.894427
13	4	1.746032	6.984126
14	2	2.116514	4.233028
15	4	2.466161	9.864644
16	2	2.618034	5.236068
17	4	2.466161	9.864644
18	2	2.116514	4.233028
19	4	1.746032	6.984126
20	2	1.447214	2.894427
21	4	1.235729	4.942917
22	2	1.099516	2.199032
23	4	1.024089	4.096357
24	2	1	2
25	4	1.024089	4.096357
26	2	1.099516	2.199032
27	4	1.235729	4.942917
28	2	1.447214	2.894427
29	4	1.746032	6.984126
30	2	2.116514	4.233028
31	4	2.466161	9.864644
32	1	2.618034	2.618034

Reference Circle

Serial	xc	yc	x	y
0	1.27202	0	1.618034	0
1	1.247578	0.248159	1.540227	0.30637
2	1.175193	0.486781	1.344083	0.556737
3	1.057646	0.706696	1.098683	0.734117
4	0.899454	0.899454	0.850651	0.850651
5	0.706696	1.057646	0.617591	0.92429
6	0.486781	1.175193	0.401273	0.96876
7	0.248159	1.247578	0.197426	0.992528
8	7.79E-17	1.27202	6.13E-17	1
9	-0.24816	1.247578	-0.19743	0.992528
10	-0.48678	1.175193	-0.40127	0.96876
11	-0.7067	1.057646	-0.61759	0.92429
12	-0.89945	0.899454	-0.85065	0.850651
13	-1.05765	0.706696	-1.09868	0.734117
14	-1.17519	0.486781	-1.34408	0.556737
15	-1.24758	0.248159	-1.54023	0.30637
16	-1.27202	1.56E-16	-1.61803	1.98E-16
17	-1.24758	-0.24816	-1.54023	-0.30637
18	-1.17519	-0.48678	-1.34408	-0.55674
19	-1.05765	-0.7067	-1.09868	-0.73412
20	-0.89945	-0.89945	-0.85065	-0.85065
21	-0.7067	-1.05765	-0.61759	-0.92429
22	-0.48678	-1.17519	-0.40127	-0.96876
23	-0.24816	-1.24758	-0.19743	-0.99253
24	-2.3E-16	-1.27202	-1.8E-16	-1
25	0.248159	-1.24758	0.197426	-0.99253
26	0.486781	-1.17519	0.401273	-0.96876
27	0.706696	-1.05765	0.617591	-0.92429
28	0.899454	-0.89945	0.850651	-0.85065
29	1.057646	-0.7067	1.098683	-0.73412
30	1.175193	-0.48678	1.344083	-0.55674
31	1.247578	-0.24816	1.540227	-0.30637
32	1.27202	-3.1E-16	1.618034	-4E-16

Table 2.7
The 32 Interval Calculations
for a
1:Φ Ellipse

The relative merits of Sector Summation; Simpson; Meister's Rule and Mean Radius methods is already clear, but Table 2.8 shows the comparisons for the 1024-interval 1:Φ Ellipse. Meister's Rule and the Summation of Sectors are still tantamount, but Simpson Integration is eleven orders of magnitude better than either.

The error of Mean Radius estimation is 2.78%.

Number of Radial Intervals	1024
Number of Perimeter Points	1025
Range (2π)	6.283185307179590
$d\theta$	0.006135923151543
Sine of $d\theta$	0.006135884649154
Φ	1.618033988749890
a	1.618033988749890
b	1
ecc	0.786151377757423
E_{ecc} CEI 2nd Integral	1.288568294506470
θ_0	0.000000000000000
θ_n	6.283185307179590
Simpson Interval, h	0.006135923151543
Mean Radius	1.254205855403070
$\pi^*\mu_r^2$	4.941826804647460
$A_{Meister}$	5.083159834667740
$A_{ellipse}$	5.083203692315260
$A_{sectors}$	5.083159834667750
$A_{Simpson}$	5.083203692315260
PSD($A_{ellipse}$,$A_{sectors}$)	0.000862795397696
PSD($A_{ellipse}$,$A_{Simpson}$)	0.000000000000070
PSD($A_{ellipse}$,$A_{Meister}$)	0.000862795397888
PSD($A_{ellipse}$,$A_{\pi r^2}$)	2.781255606214040
$C_{ellipse}$	8.339789189347810
C_{approx}	8.339765962128610
PSD($C_{ellipse}$,C_{approx})	0.000278510867342
Radius of Reference Circle	1.272019649514070
Area of Reference Circle	5.083203692315260

Table 2.8
Comparative Table of the Inputs and Outcomes of a 1024-Interval Quadrature of a 1:Φ Ellipse

The Fiducial Bean Curve

The planar plot of a Bean Curve is reniform. I agree with those who think it looks like the profile of a lima bean.

The bean curve is described by this polar equation:-

$$r_i = a[(\sin\theta_i)^3 + (\cos\theta_i)^3]$$

Equation 2.36

This equation traces the red curve of Figure 2.4:-

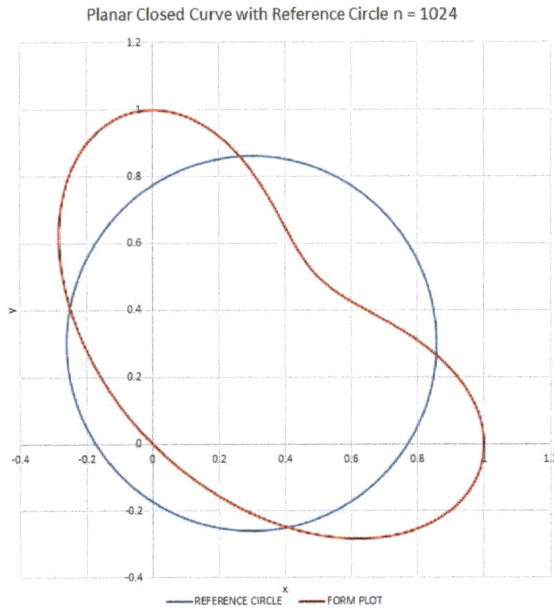

Figure 2.4
The Fiducial Bean Curve

Table 2.9 shows the comparative input and outcome data for the simple quadratures of a bean curve.

The relative merits of Sector Summation; Simpson; Meister's Rule and Mean Radius methods are as clear as for the Ellipse, and Table 2.9 shows the comparisons for the 1024-interval a=1 Bean Curve. Meister's Rule and the Summation of Sectors are still tantamount, but Simpson Integration is eleven orders of magnitude better than either.

The error of Mean Radius estimation is 42.47%.

For the Bean Curve, the use of simple Boundary Chord Summation is a highly satisfactory estimator of perimeter, showing a PSD of only 0.00027%.

Number of Radial Intervals	1024
Number of Perimeter Points	1025
Bean Curve Axix, a	1
Bean Curve Fiducial Area, A	0.981747704246810
Bean Curve Fiducial Perimeter, s	3.931702799441930
Geometric Centroid, x bar	0.3
Geometric Centraid, y bar	0.3
Range (π)	3.141592653589790
$d\theta$	0.003067961575771
Sine of $d\theta$	0.003067956762966
θ_0	0.000000000000000
θ_n	3.141592653589790
Simpson Interval, h	0.003067961575771
Mean Radius	0.423999119938293
$\pi^*\mu_r^2$	0.564780616347697
$A_{Meister}$	0.981737847663293
$A_{sectors}$	0.981737847663292
$A_{Simpson}$	0.981747704246810
PSD($A_{fiducial}$,$A_{sectors}$)	0.001003983352972
PSD($A_{fiduicial}$,$A_{Simpson}$)	0.000000000000045
PSD($A_{ellipse}$,$A_{Meister}$)	0.001003983352791
PSD($A_{ellipse}$,$A_{\pi r^2}$)	42.471918813297100
$C_{fiducial}$	3.931702799441930
C_{approx}	3.931692188440450
PSD($C_{fiducial}$,C_{approx})	0.000269883102069
Radius of Reference Circle	0.559016994374947
Area of Reference Circle	0.981747704246811
Range of Reference Circle (2π)	6.283185307179590
$d\theta_c$	0.006135923151543

Table 2.9
Comparative Table of the Inputs and Outcomes of a 1024-Interval Quadrature of a Bean Curve that has a = 1

The Superiority of Simpson's Rule

Table 2.10 provides a summary conspectus of variables for the Fiducial Ellipse and $n = 2^m$ for $m = 3, 4, 5 \ldots 10$:-

Number of Radial Intervals	8	16	32	64	128	256	512	1024
Log_2(Number of Radial Intervals)	3	4	5	6	7	8	9	10
Number of Perimeter Points	9	17	33	65	129	257	513	1025
Range (2π)	6.283185	6.283185	6.283185	6.283185	6.283185	6.283185	6.283185	6.283185
$d\theta$	0.785398	0.392699	0.19635	0.098175	0.049087	0.024544	0.012272	0.006136
Sine of $d\theta$	0.707107	0.382683	0.19509	0.098017	0.049068	0.024541	0.012272	0.006136
Φ	1.618034	1.618034	1.618034	1.618034	1.618034	1.618034	1.618034	1.618034
a	1.618034	1.618034	1.618034	1.618034	1.618034	1.618034	1.618034	1.618034
b	1	1	1	1	1	1	1	1
ecc	0.786151	0.786151	0.786151	0.786151	0.786151	0.786151	0.786151	0.786151
E_{ecc} CEI 2nd Integral	1.288568	1.288568	1.288568	1.288568	1.288568	1.288568	1.288568	1.288568
θ_0	0	0	0	0	0	0	0	0
θ_n	6.283185	6.283185	6.283185	6.283185	6.283185	6.283185	6.283185	6.283185
Simpson Interval, h	0.785398	0.392699	0.19635	0.098175	0.049087	0.024544	0.012272	0.006136
$A_{ellipse}$	5.083204	5.083204	5.083204	5.083204	5.083204	5.083204	5.083204	5.083204
$A_{sectors}$	4.454065	4.909164	5.038654	5.071999	5.080398	5.082502	5.083028	5.08316
$A_{simpson}$	4.925435	5.072777	5.083171	5.083204	5.083204	5.083204	5.083204	5.083204
PSD($A_{ellipse}$,$A_{sectors}$)	12.37681	3.423824	0.876409	0.220429	0.055191	0.013803	0.003451	0.000863
PSD($A_{ellipse}$,$A_{simpson}$)	3.103721	0.205114	0.000643	6.2E-09	1.75E-14	-3.5E-14	1.75E-14	6.99E-14
$C_{ellipse}$	8.339789	8.339789	8.339789	8.339789	8.339789	8.339789	8.339789	8.339789
C_{approx}	8.037195	8.247799	8.316155	8.333852	8.338303	8.339418	8.339696	8.339766
PSD($C_{ellipse}$,C_{approx})	3.628318	1.103025	0.283387	0.071186	0.017818	0.004456	0.001114	0.000279
Radius of Reference Circle	1.27202	1.27202	1.27202	1.27202	1.27202	1.27202	1.271998	1.27202
Area of Reference Circle	5.083204	5.083204	5.083204	5.083204	5.083204	5.083204	5.083028	5.083204

Table 2.10
Input and Outcome Variables
for the 1:Φ Ellipse
with 8 to 1024 Intervals

You can see that whilst Sector Summation (and Meister's Rule) quadratures improve linearly with the geometrical increase of boundary interval count the accuracy of Simpson's Rule increases exponentially.

This is illustrated by Figure 2.5:-

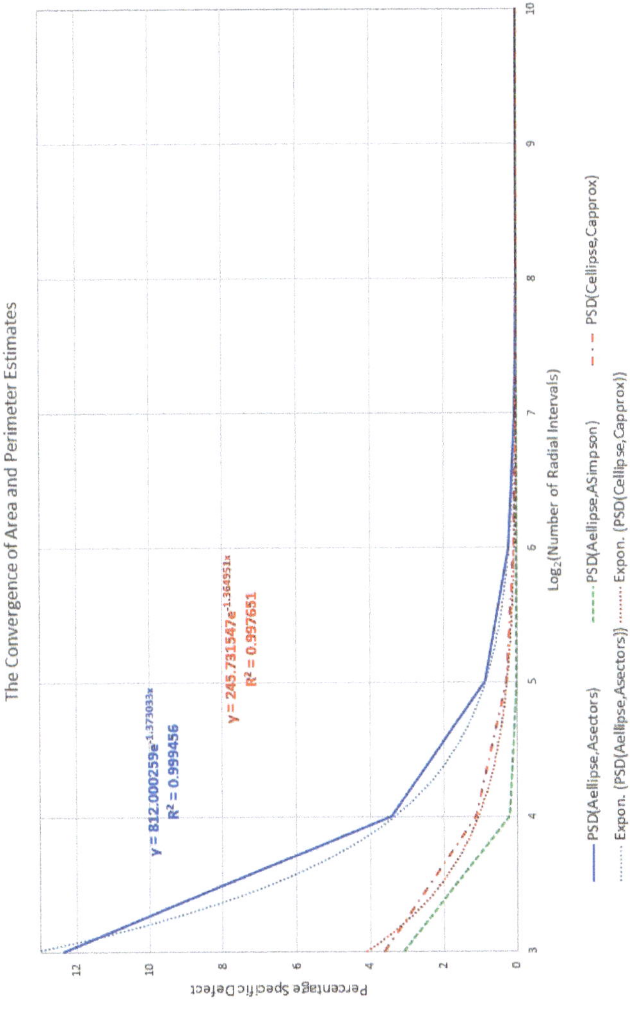

Figure 2.5
The Behavior of PSD with n the Number of Intervals for Sector and Simpson Area Estimates and for Segmental Perimeter Estimates

The blue line regression is for Sectoral estimates and the green dashed line is Simpson estimate behavior. Meanwhile the red regression shows the fidelity of Segmental perimeter estimates.

To conceive a clearer appreciation of the gross superiority of Simpson's Rule in simple quadrature it is helpful to take logarithms as in Table 2.11, and examine Figure 2.6

As you can see, the accuracy of Sector Summation converges linearly, whereas Simpson's Rule converges quadratically.

Number of Radial Intervals	8	16	32	64	128	256	512	1024
Log_2(Number of Radial Intervals)	3	4	5	6	7	8	9	10
Number of Perimeter Points	9	17	33	65	129	257	513	1025
Range (2π)	6.283185	6.283185	6.283185	6.283185	6.283185	6.283185	6.283185	6.283185
$d\theta$	0.785398	0.392699	0.19635	0.098175	0.049087	0.024544	0.012272	0.006136
Sine of $d\theta$	0.707107	0.382683	0.19509	0.098017	0.049068	0.024541	0.012272	0.006136
Φ	1.618034	1.618034	1.618034	1.618034	1.618034	1.618034	1.618034	1.618034
a	1.618034	1.618034	1.618034	1.618034	1.618034	1.618034	1.618034	1.618034
b	1	1	1	1	1	1	1	1
ecc	0.786151	0.786151	0.786151	0.786151	0.786151	0.786151	0.786151	0.786151
E_{ecc} CEI 2nd Integral	1.288568	1.288568	1.288568	1.288568	1.288568	1.288568	1.288568	1.288568
θ_0	0	0	0	0	0	0	0	0
θ_n	6.283185	6.283185	6.283185	6.283185	6.283185	6.283185	6.283185	6.283185
Simpson Interval, h	0.785398	0.392699	0.19635	0.098175	0.049087	0.024544	0.012272	0.006136
$A_{ellipse}$	5.083204	5.083204	5.083204	5.083204	5.083204	5.083204	5.083204	5.083204
$A_{sectors}$	4.454065	4.909164	5.038654	5.071999	5.080398	5.082502	5.083028	5.08316
$A_{Simpson}$	4.925435	5.072777	5.083171	5.083204	5.083204	5.083204	5.083204	5.083204
PSD($A_{ellipse}$,$A_{sectors}$)	12.37681	3.423824	0.876409	0.220429	0.055191	0.013803	0.003451	0.000863
PSD($A_{ellipse}$,$A_{Simpson}$)	3.103721	0.205114	0.000643	6.2E-09	1.75E-14	-3.5E-14	1.75E-14	6.99E-14
$C_{ellipse}$	8.339789	8.339789	8.339789	8.339789	8.339789	8.339789	8.339789	8.339789
C_{approx}	8.037195	8.247799	8.316155	8.333852	8.338303	8.339418	8.339696	8.339766
PSD($C_{ellipse}$,C_{approx})	3.628318	1.103025	0.283387	0.071186	0.017818	0.004456	0.001114	0.000279
Radius of Reference Circle	1.27202	1.27202	1.27202	1.27202	1.27202	1.27202	1.271998	1.27202
Area of Reference Circle	5.083204	5.083204	5.083204	5.083204	5.083204	5.083204	5.083028	5.083204
Log_2(Number of Radial Intervals)	3	4	5	6	7	8	9	10
PSD($A_{ellipse}$,$A_{sectors}$)	12.37681	3.423824	0.876409	0.220429	0.055191	0.013803	0.003451	0.000863
PSD($A_{ellipse}$,$A_{simpson}$)	3.103721	0.205114	0.000643	6.2E-09	1.75E-14	-3.5E-14	1.75E-14	6.99E-14
PSD($C_{ellipse}$,C_{approx})	3.628318	1.103025	0.283387	0.071186	0.017818	0.004456	0.001114	0.000279
Log_2(Number of Radial Intervals)	3	4	5	6	7	8	9	10
log_e(PSD($A_{ellipse}$,$A_{sectors}$))	2.515824	1.230758	-0.13192	-1.51218	-2.89695	-4.28286	-5.66906	-7.05533
log_e(PSD($A_{ellipse}$,$A_{simpson}$))	1.132602	-1.58419	-7.34941	-18.8985	-31.6781	-30.985	-31.6781	-30.2918
log_e(PSD($C_{ellipse}$,C_{approx}))	1.288769	0.098057	-1.26094	-2.64246	-4.02756	-5.41356	-6.79978	-8.18605
Log_2(Number of Radial Intervals)	3	4	5	6	7	8	9	10
log_{10}PSD($A_{ellipse}$,$A_{sectors}$)	1.092609	0.534511	-0.05729	-0.65673	-1.25813	-1.86002	-2.46204	-3.06409
log_{10}(PSD($A_{ellipse}$,$A_{simpson}$))	0.491883	-0.688	-3.19181	-8.2075	-13.7576	-13.4566	-13.7576	-13.1556

Table 2.11
Input and Outcome Variables for the 1:Φ Ellipse with Plotting Logarithms

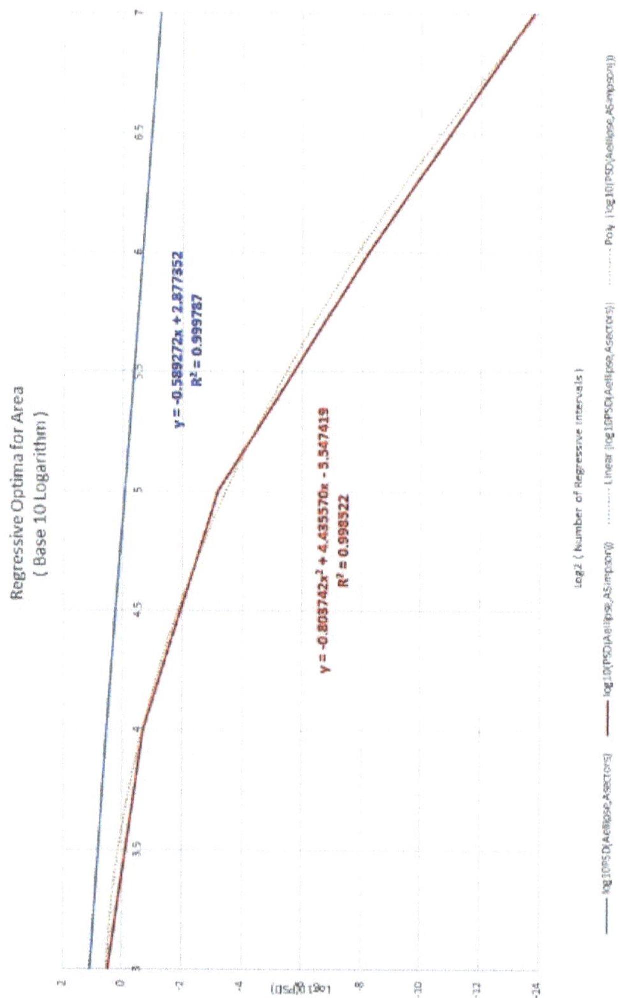

Figure 2.6
The Logarithms of the Behavior of PSD with n the Number of Intervals for Sector and Simpson Area Estimates and for Segmental Perimeter Estimates

Areal Equivalencies:- Sectors' Sum and Simpson's Rule

It is manifest, especially for ideal figures such as the Ellipse and the Bean Curve, that the Sum of Polar Sectors is about the same as the Simpson's Rule Integral, and also at least approximately the same as the Meister's Rule Areal Estimate.

After all, all three methods ostensibly measure the size of an objective area.

We may explore these equivalencies further.

To have a fiducial area to act as a point of reference we shall adopt the Area of a Φ:1 Ellipse, $A_{ellipse}$, such that:-

$$A_{fido} = A_{ellipse} = \pi ab = 5.08320369231526$$

Equation 2.37

For the sake of numerical integrities we shall assume that n = 1024 which is "sufficiently large", and adopt equivalency as a postulate, that is to say:-

$$A_{fido} = A_{sector}$$

Equation 2.38

and:-

$$A_{sector} \equiv A_{Simpson} \equiv A_{Meiser}$$

Equation 2.39

and for the Area as the Sum of Sectors, A_{sector}:-

$$A_{sector} = \sum_{k=0}^{n-1} dA_k = \sum_{k=0}^{n-1} \frac{1}{2}.r_k.r_{k+1}.\sin\left(\frac{2\pi}{n}\right)$$
$$= 5.08315983466775$$

Equation 2.40

Meanwhile, our Polar form of Simpson's Rule is:-

$$A_{Simpson} = \frac{1}{6}.\frac{2\pi}{n}.\left\{r_0{}^2 + \sum_{kk=1}^{n-1} [3 + (-1)^{kk-1}].r_{kk}{}^2 + r_n{}^2\right\}$$
$$= 5.08320369231526$$

Equation 2.41

k and kk are arbitrary counters.

The relevant Percentage Specific Defects are:-

$$PSD(A_{ellipse}, A_{sector}) = 0.000862795397678$$

Equation 2.42a

$$PSD(A_{sector}, A_{Simpson}) = -0.000862802841814$$

Equation 2.42b

$$PSD(A_{ellipse}, A_{Simpson}) = 0.000000000000087$$

Equation 2.42c

Hence by our postulate:-

$$\sum_{k=0}^{n-1} \frac{1}{2}.r_k.r_{k+1}.\sin\left(\frac{2\pi}{n}\right)$$
$$= \frac{1}{6}.\frac{2\pi}{n}.\left\{r_0{}^2 + \sum_{kk=1}^{n-1} [3 + (-1)^{kk-1}].r_{kk}{}^2 + r_n{}^2\right\}$$

Equation 2.43

and should we desire to work in Cartesian terms we may note that *if the system is centered at the origin*, that is to say the base of each radius is at x = 0 and y = 0, then by the Pythagoras Theorem:-

$$r_k{}^2 = x_k{}^2 + y_k{}^2$$

Equation 2.44

and by suitable substitutions we may attain:-

$$\frac{1}{2}.\sin\left(\frac{2\pi}{n}\right).\sum_{k=0}^{n-1} \sqrt{(x_k{}^2 + y_k{}^2)(x_{k+1}{}^2 + y_{k+1}{}^2)}$$
$$= \frac{\pi}{3n}\left\{(x_0{}^2 + y_0{}^2) + \sum_{kk=1}^{n-1} [3 + (-1)^{kk-1}].(x_{kk}{}^2 + y_{kk}{}^2) + (x_n{}^2 + y_n{}^2)\right\}$$

Equation 2.45

In a well-measured closed figure it is likely bordering upon certain that r_0 will be the same as r_n and indeed that for an ideal ellipse:-

$$r_0 = r_n = a$$

Equation 2.46

If the Cartesian Increments dx and dy are "sufficiently small" we may substitute for the LHS:-

$$A_{sector} \approx \frac{\pi}{n}\sum_{k=0}^{n-1}\sqrt{{r_k}^2+{r_{k+1}}^2} \approx \frac{\pi}{n}\sum_{k=0}^{n-1}{r_k}^2$$

Equation 2.47

This gives rise to:-

$$PSD(A_{sector}, A_{Simpson}) = -0.00000000000021$$

Equation 2.48

It follows of course that this approximation is also valid for sufficiently small dθ.

<u>Areal Equivalencies:- Sectors' Sum and Meister's Rule</u>

From Equations 2.38 to 2.40 we may adduce a similar argument for Meister's Rule, i.e.:-

$$A_{sector} = A_{Meister}$$

Equation 2.49

or:-

$$\sum_{k=0}^{n-1} dA_k = \sum_{k=0}^{n-1} \frac{1}{2}.r_k.r_{k+1}.\sin\left(\frac{2\pi}{n}\right) = \frac{1}{2}.\sum_{i=1}^{n-1} y_i \cdot (x_{i-1} - x_{i+1})$$

Equation 2.50

where i is another arbitrary counter.

After conforming the LHS to Cartesian terms we may write:-

$$\frac{1}{2}.\sin\left(\frac{2\pi}{n}\right).\sum_{i=0}^{n-1} \sqrt{(x_i^2 + y_i^2)(x_{i+1}{}^2 + y_{i+1}{}^2)} = \frac{1}{2}.\sum_{i=1}^{n-1} y_i \cdot (x_{i-1} - x_{i+1})$$

Equation 2.51

The PSD of this is:-

$$PSD(A_{sector}, A_{Meister}) = -0.000000000000017$$

Equation 2.52

and where dx and dy are "sufficiently small":-

$$\frac{1}{2}.\sin\left(\frac{2\pi}{n}\right).\sum_{i=0}^{n-1} \sqrt{(x_i^2 + y_i^2)(dx_i{}^2 + 2.dx_i.x_i + dy_i{}^2 + 2.dy_i.y_i + x_i^2 + y_i^2)}$$

$$= \frac{1}{2}.\sum_{i=1}^{n-1} y_i \cdot (x_{i-1} - x_{i+1})$$

Equation 2.53

which by neglecting infinitesimals gives:-

$$\frac{1}{2}.\sin\left(\frac{2\pi}{n}\right).\sum_{i=0}^{n-1}(x_i{}^2 + y_i{}^2) \approx \frac{1}{2}.\sum_{i=1}^{n-1} y_i \cdot (x_{i-1} - x_{i+1})$$

Equation 2.54

and that reduces to:-

$$\frac{1}{2}.\sin\left(\frac{2\pi}{n}\right).\sum_{i=0}^{n-1} r_i{}^2 \approx \frac{1}{2}.\sum_{i=1}^{n-1} y_i \cdot (x_{i-1} - x_{i+1})$$

Equation 2.55

for which the PSD is:-

$$PSD(A_{sector}, A_{Meister}) = 0.000235305506673$$

Equation 2.56

It roughly follows that dθ is also "sufficiently small" so that the Sine approximates its argument.

Hence:-

$$\frac{\pi}{n}.\sum_{i=0}^{n-1} r_i{}^2 \approx \frac{1}{2}.\sum_{i=1}^{n-1} y_i \cdot (x_{i-1} - x_{i+1})$$

Equation 2.57

for which the PSD is:-

$$PSD(A_{sector}, A_{Meister}) = 0.000862795397608$$

Equation 2.58

PART THREE
LAKES AND THEIR AREAS

Types of Glacial Lakes

Glacial lakes are stretches of subaerial freshwater that are formed in whole or in part by the action of ice. Like any lake, glacial lakes can be a subaerial expression of the subterraneous water table, like a man-made well is.

Thick layers of ice, up to a kilometer deep or more, can form from snow accumulation at high latitudes or high altitudes.

Ice does not currently persist anywhere in the British Isles but it regularly trespasses British latitudes in cycles of tens of thousands of years. These cyclical ice ages affect Northern latitudes in Canada, the northern USA, Northern Europe and Siberia, and Southern Latitudes covering Chile, Argentina and New Zealand.

Altitudinal glaciation affects the Himalayas, the Alps, the Andes and the Rocky Mountains of the Americas, Mount Cook in New Zealand which enjoys a temperate maritime climate very like that of southern Britain, and even the lofty Kilimanjaro which is almost on the Equator.

During Ice Ages thick ice blankets the landscape slowly creeping south to melt somewhere north of the Severn-Thames latitude (~53° N). In North America the terminal latitude is about 37° N around the latitude of Maryland. This difference is because Britain is warmed by the great sea-current, the North Atlantic Drift and is therefore of similar climate to say Virginia, Pennsylvania or northern California.

Creeping ice picks up shattered rock fragments (moraine) at its contact with the ground, and uses these to erode the land surface, thus acquiring more abrasive material. When the ice retreats on the onset of warm epochs it creates lakes and other landforms by passive resistance to other geological activity. During glacial retreat, rock detritus of various kinds and grades is deposited upon the land surface as boulder clay or at ice-water margins as eskers, drumlins and moraines, which are ridges and mounds of molded sediment which can themselves dam glacial lakes.

Such glacial features, and especially lakes, possess discriminable surfaces and can therefore have such areas assessed by simple quadrature techniques.

All natural lakes have depths that are variable to some extent and these can be plumbed with a lead weight on a string to measure a point depth, or by using modern electro-acoustical techniques. This will generate a series of point readings in meters that can be averaged to give a Mean Depth and multiplied by surface area to estimate the volumetric capacity.

Alternatively, or additionally, the point depths may be interpolated with fitted line contours called isobaths which are directly comparable to contours on land maps.

It is therefore clear that isobaths are surfaces of equal depth which may mathematically be nested to generate a three-dimensional volume. The mathematical art of estimating the volumes of objects is called Cubature, but cubature is beyond the scope of this book.

In all such exercises it must be born in mind that the depth and area of even pristine lakes varies with weather and season and that many British lakes are modified by their use as reservoirs: For industrial and domestic water supply; for canal maintainence and in highland areas for hydro-electric power. Southern reservoirs often vanish altogether during summer droughts. Also, the beds of many British lakes comprise an indeterminate quicksand, sludge or organic filth which complicates assessment of useful water capacities.

There are three key varieties of glacial lake in Britain and Ireland: The Finger or Ribbon lake; the Tarn; and the Kettle or Kettle Hole.

(A) **Finger Lakes**

Finger lakes are often kilometers long and around a kilometer wide with depths of up to 300 meters. Indeed, finger lakes near coastal outlets often have floors below sea level, and where the downstream (proximal) end is breached to the sea may form tidal inlets called fiords.

Finger Lakes typically have braided river deltas at their upstream (distal) ends and simple proximal outflows.

Finger or Ribbon lakes are almost always closely-confined in U-shaped Valleys between high mountains and are characteristic of the Scottish Highlands and the Cumbrian Lake District.

The sides of finger lakes are precipitous and their floors flat. The sides are of bedrock and the floors silt-covered.

Discussions of finger lakes must carefully discriminate rift-valley lakes which are superficially the same but are formed, at least in part, between parallel geological fractures (faults) in the Earth's Crust.

For example, Loch Ness in Scotland, which contains more fresh water, 7.452×10^9 cubic meters or about 1.8 cubic miles, than all the lakes of England and Wales put together, but is more of a rift-valley formation like Lake Nyasa in Africa than like Loch Earn which is also in the Scottish Highlands.

Figure 3.1 is the bathymetric plan of a typical glacial finger lake including a selected scaled cross-section:-

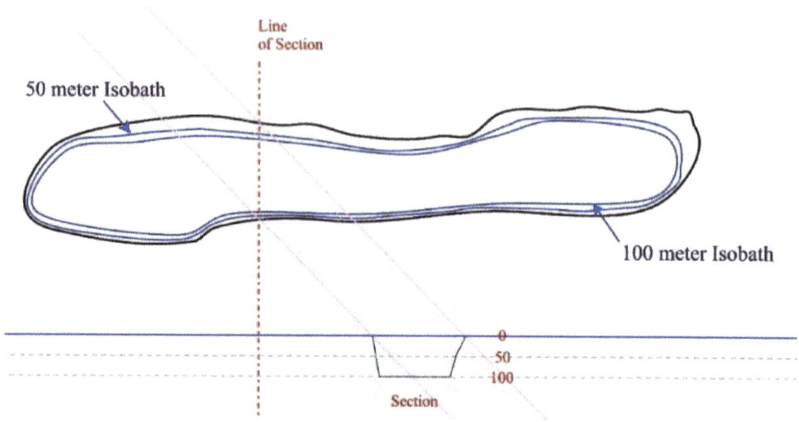

Figure 3.1
An Typical Glacial Finger or Ribbon Lake

The National Library of Scotland keeps a set of some hundreds of bathymetric maps from "*Bathymetric Survey of the Freshwater Lochs of Scotland*" of Scottish lakes surveyed by Sir John Murray FRS and Mr L Pullar FRSE during the years 1900-1905AD which map point sounding transects and isobaths, presumably made from a rowing boat.

This collection includes surveys of Loch Ness and Loch Earn reproduced below:-

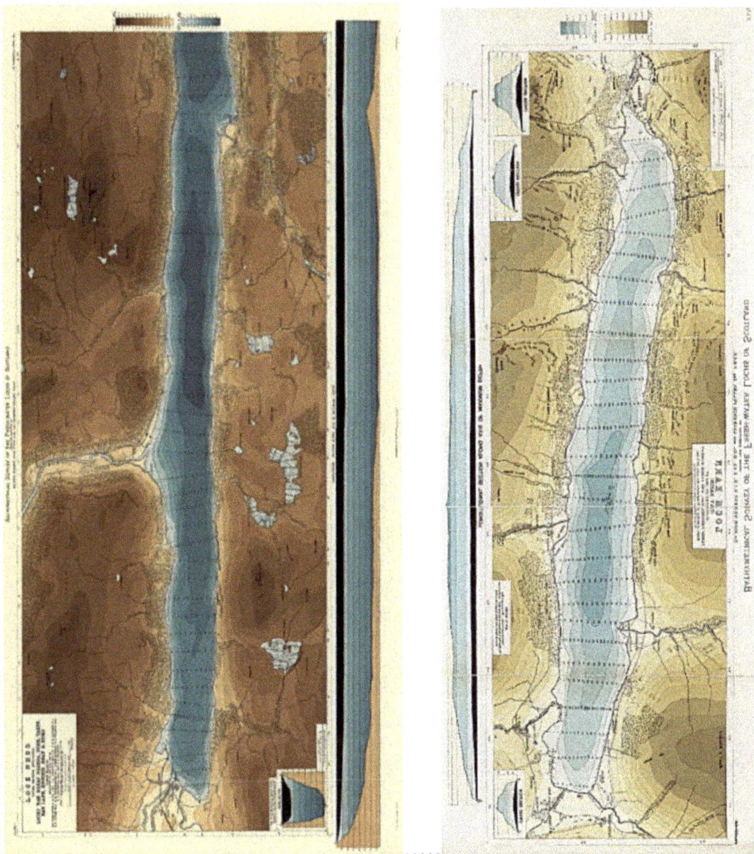

Figure 3.2
Edwardian Bathymetric Maps of Lochs Ness and Earn
By Sir John Murray FRS and Mr L Pullar FRSE
(NLS Map Archive)

(B) Tarns

Tarns are roughly oval, relatively shallow lakes that form by intense ice gouging in the corrie, sometimes called a cirque or cwm, which is an ice-amphitheater at the heads of glaciers.

Tarns are often quasi-circular, but teardrop shaped and heart shaped examples are frequently found.

These terminal hollows form on the face of a pyramidal rock peak, usually a high mountain. These terminal hollows are the head of a valley glacier where compact snow and ice falls from the mountain face, itself weakened by the diurnal freeze-thaw spallation of rock. The ice compresses and erodes the foot of the rock face forming a deep depression that fills with water when the ice melts. The water-filled depression is broadly saucer shaped, typically thirty meters deep, but deepest at the distal end below the heading cliff face. The proximal shore is often stony and gently-shelving, or locally sandy, comprising moraine rock-debris which enhances the total depth of the lake.

The floor is of bedrock or scree.

Llyn Cau on the East face of the Cader Idris in Wales is a classic primeval tarn, famous and much depicted not only by photographers but also the Romantic landscape artists of the past.

There are many tarns in the Cumbrian Lake District of North-West England, a Romantic and modern tourist resort, a National Park celebrated for poets and artists such as Wordsworth, Turner and Beatrix Potter.

But many of the Cumbrian lakes including the tarns have been interfered with: The tarns for hydro power for mining and the larger valley ribbon lakes for long-distance city water supply.

Tarns occur, however, worldwide, wherever there are bare rock mountains which are, or were, glaciated.

Figure 3.3 schematises the plan and proximal cross-section of a typical mountain tarn:-

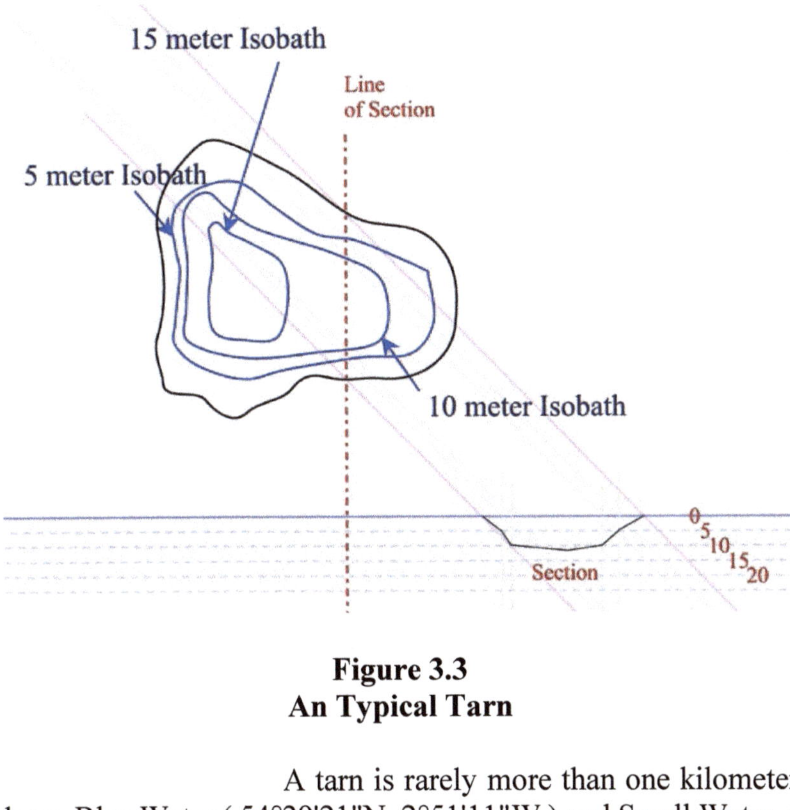

Figure 3.3
An Typical Tarn

A tarn is rarely more than one kilometer long. Blea Water (54°29'21"N, 2°51'11"W) and Small Water (54°28'58"N, 2°50'34"W) are two classic quasi-circular tarns that sit in textbook cirques within one kilometer of one another across an eastern arête of Racecourse Hill in the Cumbrian Lake District. Both are near pristine, Blea Water being raised a little by a low weir.

Blea Water has a surface area of 17 hectares and at 63.09 meters is the deepest tarn in the Cumbrian Lake District.

(C) Kettles or Kettle Holes

Kettles are lowland ice-formed lakes formed by passive ice-melting rather than attrition. Whereas ribbon lakes and tarns are erosional highland forms, kettles are lowland and depositional. Kettle holes seldom have significant surface drainage.

There are two species of kettle hole lake discriminated by their Exposed or Concealed mode of genesis.

An Exposed Kettle forms when a fractured mass of ice about one kilometer cube detaches from a retreating glacier to sit subaerially upon a plain of soft sediment. As it sits it sinks into the detritus around its foot whilst further glacial debris, typically a boulder clay, settles about it. And of course it melts as the glaciation retreats in a warming climate. It leaves a shallow water-filled depression often with a distinct shore berm: A kettle hole.

Figure 3.4 is a diagrammatic section of a Exposed remnant ice block forming its kettle whilst Figure 3.5 is a diagrammatic section of a buried remnant ice block such as that which may from a Concealed Kettle Hole as it melts.

A Concealed Kettle is actually buried ice within the deposited strata of glacial outwash debris. It too may be a block one kilometer thick but with its upper surface not buried that far. It eventually melts and the supercumbent sediment collapses into the void forming a shear-sided subaerial pit that fills with water to form a Concealed Kettle.

Figure 3.6 is the bathymetric plan of a typical glacial kettle hole including a selected scaled cross-section.

When I describe and discriminate the natures and origins of glacial lakes there is an inevitable element of caricature, as there must be in any relation of the visage of the Earth.

No geologist was present to witness the birth of lakes and so science progresses tentatively and uncertainly, clinging to The Doctrine of Uniformity, in which processes at work today are supposed to be similar to the processes that sculpted the past.

As with men and women we may think we discern particular types as we might the distinct complexions of different breeds of co-specific animals, but there is vast variation in the shape of lakes as exemplified by the bathymetric charts of White Mere and Cole Mere in the Salopian Lake District, a small swarm of kettle holes.

These are shown in Figure 2.7 and I have abstracted them from https://fishing-app.gpsnauticalcharts.com/i-boating-fishing-web-app/fishing-marine-charts-navigation.html?title=White+Mere+boating+app#13.18/52.9019/-2.8595, presented by gpsnauticalcharts.com

The extreme asymmetry of the two lake beds is clear. The isobaths appear to be at meter intervals and spot depths in meters are annotated in bold blue type.

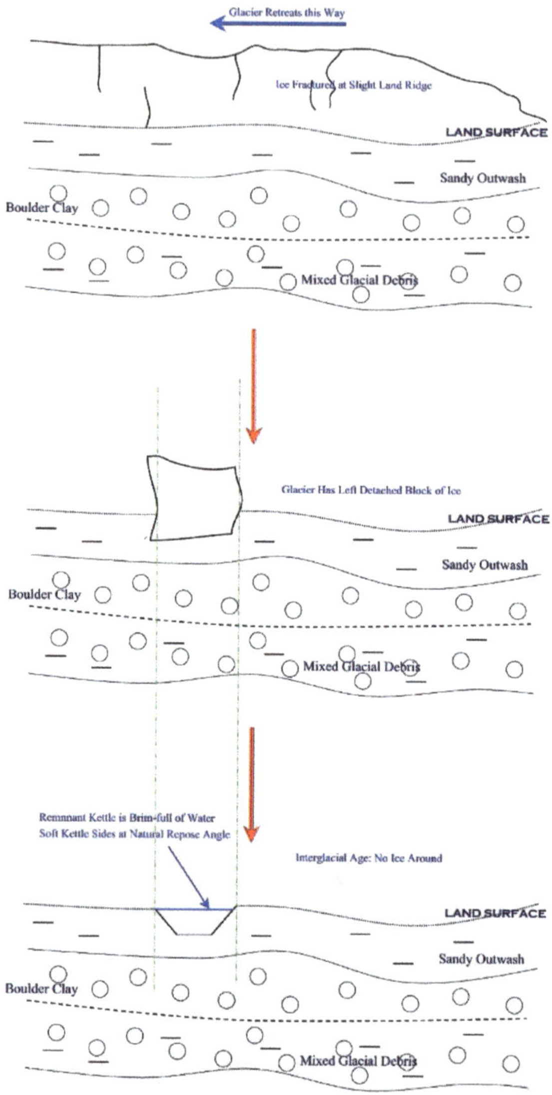

Figure 3.4
The Formation of an Exposed Kettle Hole

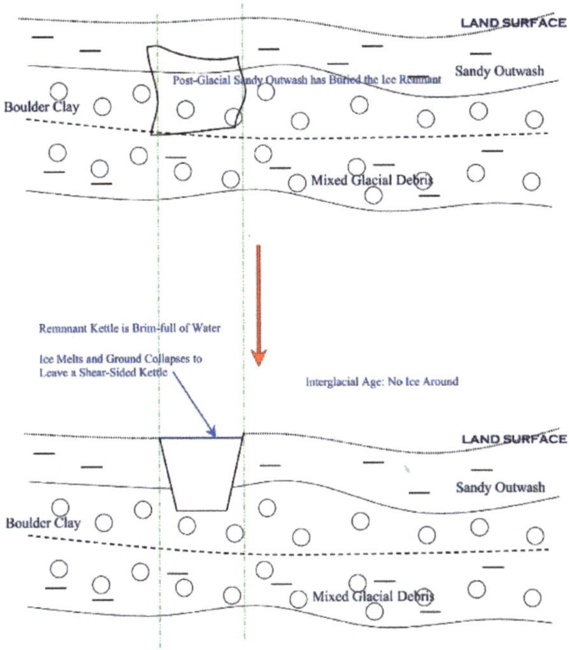

Figure 3.5
The Formation of a Concealed Kettle Hole

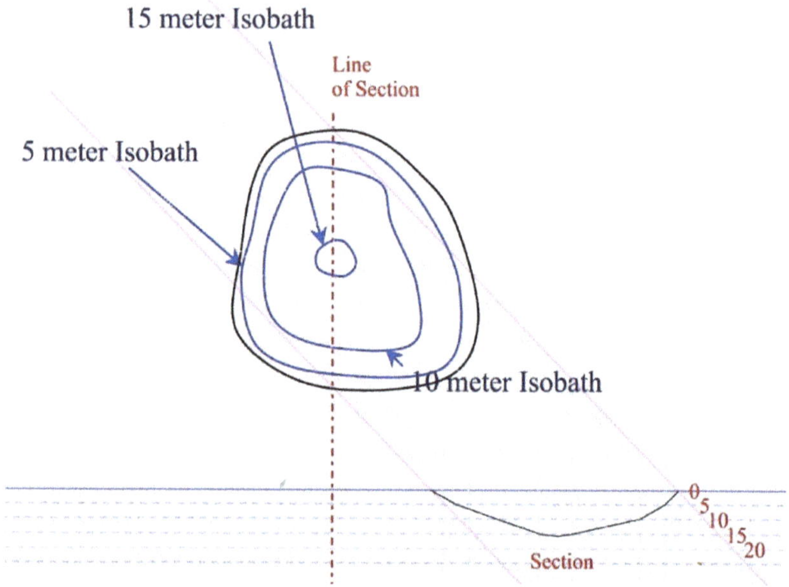

Figure 3.6
A Typical Kettle Hole

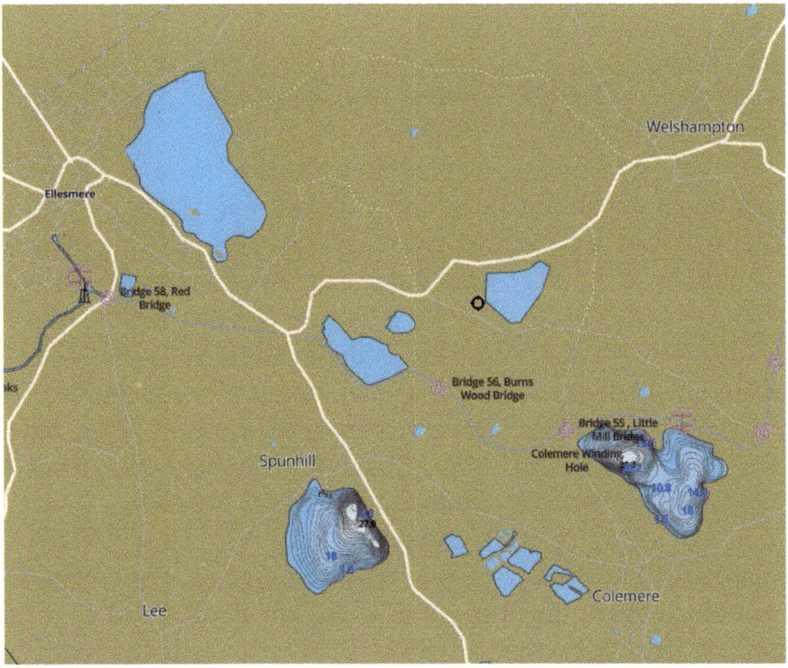

Figure 3.7
Bathymetric Charts of
White Mere and Cole Mere

Some Lakes and their Areas

Table 3.1 presents the key statistics of a few well-known British lakes.

Bomere, a few miles south of Shrewsbury is a great anomaly. It is England's only oligotrophic natural lowland lake. It is said to be crystal-clear, and though I have not visited it certainly seems very limpid from the available photographs. Most natural oligotrophic lakes are in the far highlands of Scotland or Northern England and oligotrophy typifies tarns, which however often harbour rare or unique fish populations.

Talkin Tarn is an isolated lowland kettle hole lake in East Cumberland, surrounded by Scots Pines and delightful as a family swimming pond in fine weather.

Blea Water, Small Water, Llyn Cau and Glaslyn (Snowdon) are all classical bare-mountain cirque tarns whilst Innominate Tarn and Brothers Water are perched lakes high upon impermeable strata, in both cases I believe the Borrowdale Volcanics.

Comber Mere is a probable Cheshire kettle hole of extended valley-bound shape, much modified to be an aristocratic boating lake.

Loch Ness and Wastwater are both, at least in part, glacial ribbon lakes of great depth and precipitous parallel sides.

The Lake of Menteith, often reconned a lowland kettle hole, is more probably formed by extra-terrestrial ice impacting at great velocity to generate a crater swarm.

Ellesmere, White Mere, Blake Mere, Kettle Mere, Cole Mere and Newton Mere are all classical kettle holes of the Salopian Lake District and we shall compute the surface areas of some of these.

Quadratures of Ellesmere the lake should be adjusted for the presence of Moscow Island.

The Salopian Lake District

The Salopian, or Shropshire, Lake District is a cluster of nine substantial flooded kettle holes and a swarm of lesser ponds centered about Ellesmere Town (52°54'27"N, 2°53'22"W) in the extreme North of Shropshire near the border with Cheshire county, in Western central England.

It is a lowland deciduously-wooded rural area with plentiful pasture. Ellesmere Town was formerly a major rail and canal center, fashionable as a boating resort, but largely forsaken by tourists since the First World War.

As a young man I greatly enjoyed rowing but seemed to get little opportunity to do so. I last rowed a dinghy on Ellesmere, once in the Summer of 1987 and again in August 1988. I drove from Featherstone, North of Wolverhampton.

As of January 2026 it is no longer possible to hire a boat on Ellesmere but you can use your own non-powered boat with a valid permit.

There is a barely-perceptible ridge of land underlain by Triassic Sandstone striking in a Northeasterly direction between Tetchill and Hanmer and as blanket ice crept back to the North-West, from the Mercian Plateau to the Cheshire Plain, it fractured over this bluff. Fracture blocks remained static and sank to form kettles in the district East of Ellesmere Town.

Figure 3.8 is a general OS 1:25000 map of the area taken from The National Library of Scotland online map archive. The graticules are at one kilometer intervals. The map dates to circa 1950AD.

Lake Name	Type	LOCATION (Nominal Center) Lat °	'	"	Long °	'	"	OS Grid Ref	Number of Islands	Reservoir?	OS Area	Units	OS Area (m^2)	DEFRA Water Body ID	DEFRA Area (m^2)	DEFRA Mean Depth (m)	Maximum Depth (m)	Indicated Volume (m^3)	Indicated Mass (tonnes)	Notes
Bomere	Kettle?	52	40	5	2	44	32	SJ4991808066	0	0	23.756	Acres	96137							
Talkin Tarn	Kettle	54	55	19	2	42	36	NY5449958829	1	0	64.773	Acres	262127							
Llyn Cau	Tarn	52	41	37	3	54	10	SH7151012335	0	0	15	Hectares	150000							
Blea Water	Tarn	54	29	21	2	51	10	NY4485810732	0	0	17	Hectares	170000							Racecourse Hill, Mardale Forest
Small Water	Tarn	54	28	58	2	50	33	NY4551210014	0	0	4	Hectares	40000							Racecourse Hill, Mardale Forest
Innominate (Loaf) Tarn	Tarn	54	30	23	3	14	36	NY1959013042	4	0	0.45	Hectares	45000							
Glaslyn	Tarn	53	4	14	4	3	58	SH6167354564	0	0	10	Hectares	100000							Snowdon
Wastwater	Ribbon	54	26	30	3	17	30	NY1624005923	0	1	716	Acres	2897552							1.08 $mile^2$?
Brothers Water		54	30	23	2	55	26	NY4028312709	0	0	49.02	Acres	198377							or 47 acres ?
Semer Water	Kettle	54	16	50	2	7	30	SD9186287237	0	0	84.491	Acres	341923							71 acres ?
Loch Ness	Ribbon-Rift Hybrid	57	18	0	4	27	0	NH5240025900	0	0	21.8	sq.miles	56461741							One Tiny Crannog
Loch Earn	Ribbon	56	23	14	4	12	9	NN6403723835	0	0	946.7	Hectares	946700							One Tiny Crannog
Rydal Water	Twin Kettle	54	26	50	2	59	43	NY3557606207	4	0	77.422	Acres	313316	GB31229197	300000	5.30	19.81	1590000	1589205	Heron Island 0.644 acres: Little Isle 0.196 acres
Ellesmere	Kettle	52	54	31	2	53	3	SJ4063834931	1	0	113.85	Acres	460735		445000	7.60	18.80	3382000	3380309	Moscow Island is 0.26 acres
White Mere	Kettle	52	53	28	2	52	17	SJ4148132964	0	0	63.049	Acres	255150	GB30935091	235250	5.60		1317400	1316741	
Blake Mere	Kettle	52	53	58	2	52	11	SJ4159733900	0	0	20.885	Acres	84519	ID 35052	59300	5.30	13.50	312969	312813	UK Lakes Portal: Elevation 91 meters
Kettle Mere	Kettle	52	54	4	2	51	58	SJ4184634080	0	0	4.159	Acres	16831							
Cole Mere	Kettle	52	53	37	2	50	35	SJ4338933234	0	0	68.309	Acres	276437							
Newton Mere	Kettle	52	54	9	2	51	23	SJ4250334218	0	0	20.602	Acres	83373							
Comber Mere	Kettle?	52	59	51	2	36	49	SJ5819644627	6	0	130.681	Acres	528848							Nineteeth Century Islands now Encroached into Lake
Lake of Menteith	Possibly Meteoric	56	10	32	4	17	31	NN5781700438	2	0	645.499	Acres	2612244							Complex Multi-Circle Lake with Islands
Moscow Island		52	54	24	2	53	1	SJ4067634707	1	0	0.26	Acres	1052							Island of Construction Waste in Ellesmere

Table 3.1
Selected British Lakes and their Key Statistics

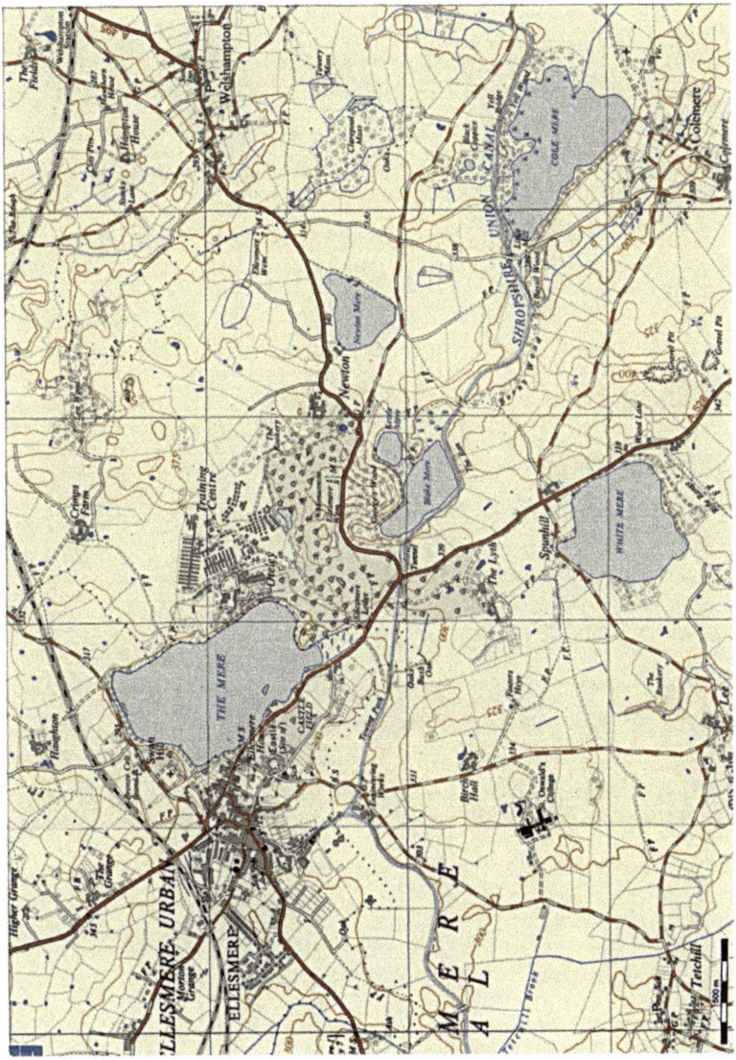

Figure 3.8
A General Map of the
Salopian Lake District

White Mere

White Mere is a vaguely circular eutrophic kettle hole lake centered about 52°53'27"N, 2°52'18"W.

Its area in acres was assessed by the Ordnance Survey to be 63.049 and this may have been taken at a Liverpool datum of 309.9 feet on 2 February 1901AD.

The maximum depth is around 8.5 meters and the DEFRA mean depth 5.6 meters as noted below.

The shores of White Mere are sheltered by mature deciduous trees and White Mere is the base of the Shropshire Sailing Club.

For the purposes of photographic conspectus, access to the Southern shore is difficult.

The UK Government Water Body ID is GB30935091.

According to the Department for Environment Food & Rural Affairs (DEFRA) at https://environment.data.gov.uk/catchment-planning/WaterBody/GB30935091?cycle=3 the mean depth of White Mere is 5.6 meters and its surface area 23.525 hectares,

According to these Attributes the Indicated Volume of White Mere is 1317400 cubic meters and given that the Density of Water at 12.08°C is 999.5 kg/m^3, the Indicated Mass of Water is about 1.317 million tonnes.

White Mere has no significant surface inlet or outlet and has high ecological phosphorous values, almost certainly due to local agricultural runoff.

Photo: James R Warren
(Olympus TG5: 16:43GMT 8 April 2026)

Figure 3.9
Sailing on White Mere captured from the Bus to Shrewsbury

I constructed a 40-interval polar rose in red and superposed it on a 1901AD 25-inch-to-the-mile (1:2534.4) OS map from the National Library of Scotland map archive at https://maps.nls.uk/geo/explore/side-by-side/#zoom=16.6&lat=52.89114&lon=-2.86933&layers=168&right=ESRIWorld.

The resulting quadrature diagram is reproduced below in Figure 3.10.

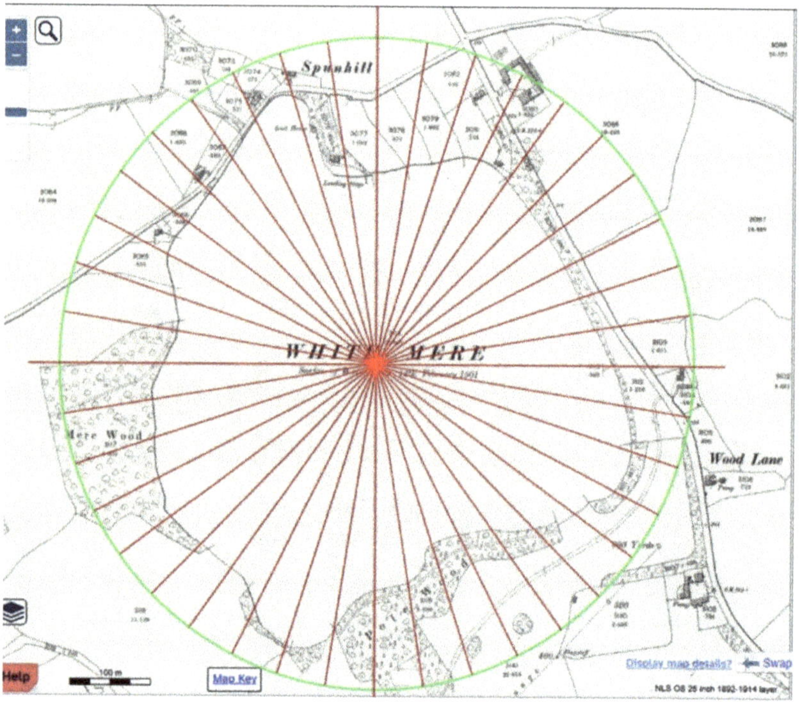

Figure 3.10
The White Mere Polar Quadrature Diagram

I processed this diagram using PhotoDraw, moving my mouse cursor by hand and reading the millimetric co-ordinates by eye. I commenced at the Easternmost intersection of the red ray with the shore and worked sinistrally recording the co-ordinates by hand until I had all forty plus one (x,y) readings.

The rose center is not exactly at the gravicenter: I shifted it slightly to the West in order to suppress ill-conditioning (by shadow) at the Spunhill shore.

These readings, and their computed angles and radii are shown in Table 3.2:-

Serial	Raw XY (pixels) X	Raw XY (pixels) Y	Centered XY (pixels) X	Centered XY (pixels) Y	Centered XY (meters) X	Centered XY (meters) Y	θ	r	r_k*r_{k+1}	dA_k	Recovered x	Recovered Y	Meister Term	dx	dy	ds
0	1571	902	605	0	286.73	0.00	0.00	286.73	77499.54	6061.80	286.73	0.00		-19.77	42.28	46.68
1	1529	811	563	91	266.82	43.13	0.16	270.29	70324.04	5500.55	266.96	42.28	1660.93	-19.51	38.12	42.82
2	1488	732	522	170	247.39	80.57	0.31	260.18	67169.97	5253.85	247.45	80.40	2969.45	-17.42	36.80	40.72
3	1451	654	485	248	229.86	117.54	0.47	258.17	66882.52	5231.37	230.03	117.20	4437.01	-20.44	35.07	40.59
4	1407	579	441	323	209.00	153.08	0.63	259.07	68067.43	5324.05	209.59	152.28	6737.09	-23.81	33.51	41.10
5	1356	508	390	394	184.83	186.73	0.79	262.74	71262.55	5573.96	185.78	185.78	9320.11	-26.36	33.64	42.74
6	1301	438	335	464	158.77	219.91	0.94	271.23	74562.44	5832.07	159.42	219.43	13380.82	-34.62	25.51	43.01
7	1229	385	263	517	124.64	245.02	1.10	274.91	69921.31	5469.05	124.80	244.94	19798.01	-46.21	-3.04	46.31
8	1130	391	164	511	77.73	242.18	1.26	254.35	61157.44	4783.57	78.60	241.90	21091.09	-40.98	-4.41	41.22
9	1046	401	80	501	37.91	237.44	1.41	240.45	54813.33	4287.35	37.61	237.49	18666.01	-37.61	-9.53	38.80
10	965	421	-1	481	-0.47	227.96	1.57	227.96	51740.93	4047.03	0.00	227.96	16668.76	-35.51	-3.79	35.71
11	891	429	-75	473	-35.55	224.17	1.73	226.97	76130.63	5954.73	-35.51	224.18	23236.02	-68.14	94.83	116.77
12	747	229	-219	673	-103.79	318.96	1.88	335.42	109556.23	8569.19	-103.65	319.00	35976.60	-44.63	-27.98	52.68
13	653	288	-313	614	-148.34	291.00	2.04	326.62	99208.25	7759.80	-148.28	291.02	21792.60	-30.25	-45.29	54.47
14	590	383	-376	519	-178.20	245.97	2.20	303.74	87640.62	6855.01	-178.53	245.73	13698.01	-25.50	-41.70	48.88
15	536	471	-430	431	-203.79	204.27	2.36	288.54	82573.86	6458.70	-204.03	204.03	10811.52	-27.49	-35.82	45.15
16	479	545	-487	357	-230.81	169.19	2.51	286.18	82493.51	6452.41	-231.52	168.21	8883.62	-25.32	-37.34	45.12
17	424	626	-542	276	-256.87	130.81	2.67	288.26	75172.18	5879.76	-256.84	130.87	2158.43	8.82	-50.28	51.05
18	443	731	-523	171	-247.87	81.04	2.83	260.78	64857.13	5072.94	-248.02	80.59	-902.41	2.37	-41.68	41.75
19	448	818	-518	84	-245.50	39.81	2.98	248.70	65064.05	5089.13	-245.64	38.91	528.94	-15.97	-38.91	42.06
20	414	901	-552	1	-261.61	0.47	3.14	261.61	71672.96	5606.06	-261.61	0.00	0.00	-8.98	-42.86	43.79
21	388	911	-578	-9	-273.93	-4.27	3.30	273.97	77399.95	6054.01	-270.59	-42.86	-303.29	1.91	-44.44	44.49
22	399	1086	-567	-184	-268.72	-87.20	3.46	282.52	84071.61	6575.85	-268.69	-87.30	475.48	3.54	-47.80	47.93
23	406	1186	-560	-284	-265.40	-134.60	3.61	297.58	92108.36	7204.46	-265.15	-135.10	2469.54	14.74	-46.83	49.10
24	437	1285	-529	-383	-250.71	-181.52	3.77	309.52	93562.41	7318.19	-250.41	-181.93	9351.97	36.66	-31.81	48.54
25	515	1353	-451	-451	-213.74	-213.74	3.93	302.28	90749.99	7098.21	-213.74	-213.74	15805.30	37.28	-29.14	47.32
26	593	1414	-373	-512	-176.78	-242.65	4.08	300.22	97295.61	7610.19	-176.46	-242.88	16179.25	29.33	-45.88	54.45
27	655	1511	-311	-609	-147.39	-288.63	4.24	324.08	114287.82	8939.28	-147.13	-288.76	19488.14	38.16	-46.63	60.25
28	734	1609	-232	-707	-109.95	-335.07	4.40	352.65	112381.39	8790.16	-108.97	-335.39	32626.24	59.12	20.64	62.62
29	860	1566	-106	-664	-50.24	-314.69	4.56	318.68	85636.09	6698.22	-49.85	-314.75	34300.20	49.85	46.03	67.85
30	963	1469	-3	-567	-1.42	-268.72	4.71	268.72	70248.90	5494.67	0.00	-268.72	24385.79	40.89	10.53	42.23
31	1051	1447	85	-545	40.28	-258.29	4.87	261.42	58339.36	4563.14	40.89	-258.20	17805.90	28.07	45.95	53.85
32	1111	1350	145	-448	68.72	-212.32	5.03	223.17	50469.48	3947.58	68.96	-212.24	13111.62	33.71	10.74	35.38
33	1181	1328	215	-426	101.90	-201.90	5.18	226.15	54412.14	4255.97	102.67	-201.50	14600.68	38.75	6.85	39.35
34	1264	1313	298	-411	141.23	-194.79	5.34	240.60	66036.28	5165.17	141.42	-194.65	17791.99	52.65	0.57	52.66
35	1376	1311	410	-409	194.31	-193.84	5.50	274.46	82715.17	6469.75	194.08	-194.08	19871.70	49.74	16.94	52.54
36	1481	1275	515	-373	244.08	-176.78	5.65	301.37	98561.64	7709.22	243.81	-177.14	17240.10	47.59	28.66	55.55
37	1581	1215	615	-313	291.47	-148.34	5.81	327.05	107051.91	8373.30	291.40	-148.48	10021.56	19.91	47.33	51.34
38	1623	1115	657	-213	311.37	-100.95	5.97	327.33	100441.96	7856.29	311.31	-101.15	1180.88	-8.23	53.15	53.78
39	1606	1000	640	-98	303.32	-46.45	6.13	306.85	88275.23	6904.64	303.08	-48.00	-1134.27	-15.40	48.00	50.41
40	1573	900	607	2	287.68	0.95	6.28	287.68			287.68	0.00				

Simpson Integration Serial	ALT_{kk}	g_{kk}	Simpson Term	Reference Circle Serial	xc	yc	x	y
0	1	82214.01	82214.01	0	284.99	0.00	286.73	0.00
1	4	73055.41	292221.65	1	281.48	44.58	266.96	42.28
2	2	67694.80	135389.59	2	271.04	88.07	247.45	80.40
3	4	66649.22	266596.89	3	253.92	129.38	230.03	117.20
4	2	67116.64	134233.28	4	230.56	167.51	209.59	152.28
5	4	69031.69	276126.77	5	201.52	201.52	185.78	185.78
6	2	73565.51	147131.02	6	167.51	230.56	159.42	219.43
7	4	75572.88	302291.50	7	129.38	253.92	124.80	244.94
8	2	64692.39	129384.78	8	88.07	271.04	78.60	241.90
9	4	57815.64	231262.55	9	44.58	281.48	37.61	237.49
10	2	51966.94	103933.87	10	0.00	284.99	0.00	227.96
11	4	51515.91	206063.66	11	-44.58	281.48	-35.51	224.18
12	2	112506.46	225012.92	12	-88.07	271.04	-103.65	319.00
13	4	106683.36	426733.45	13	-129.38	253.92	-148.28	291.02
14	2	92256.91	184513.82	14	-167.51	230.56	-178.53	245.73
15	4	83255.32	333021.27	15	-201.52	201.52	-204.03	204.03
16	2	81897.98	163795.96	16	-230.56	167.51	-231.52	168.21
17	4	83093.37	332373.49	17	-253.92	129.38	-256.84	130.87
18	2	68006.11	136012.22	18	-271.04	88.07	-248.02	80.59
19	4	61853.96	247415.83	19	-281.48	44.58	-245.64	38.91
20	2	68440.74	136881.47	20	-284.99	0.00	-261.61	0.00
21	4	75057.84	300231.35	21	-281.48	-44.58	-270.59	-42.86
22	2	79815.14	159630.29	22	-271.04	-88.07	-268.69	-87.30
23	4	88555.06	354220.26	23	-253.92	-129.38	-265.15	-135.10
24	2	95804.23	191608.45	24	-230.56	-167.51	-250.41	-181.93
25	4	91373.06	365492.24	25	-201.52	-201.52	-213.74	-213.74
26	2	90131.17	180262.35	26	-167.51	-230.56	-176.46	-242.88
27	4	105029.54	420118.15	27	-129.38	-253.92	-147.13	-288.76
28	2	124362.21	248724.42	28	-88.07	-271.04	-108.97	-335.39
29	4	101554.77	406219.09	29	-44.58	-281.48	-49.85	-314.75
30	2	72212.66	144425.33	30	0.00	-284.99	0.00	-268.72
31	4	68338.54	273354.15	31	44.58	-281.48	40.89	-258.20
32	2	49803.24	99606.48	32	88.07	-271.04	68.96	-212.24
33	4	51144.63	204578.51	33	129.38	-253.92	102.67	-201.50
34	2	57888.41	115776.82	34	167.51	-230.56	141.42	-194.65
35	4	75330.97	301323.87	35	201.52	-201.52	194.08	-194.08
36	2	90823.21	181646.41	36	230.56	-167.51	243.81	-177.14
37	4	106959.41	427837.65	37	253.92	-129.38	291.40	-148.48
38	2	107144.49	214288.99	38	271.04	-88.07	311.31	-101.15
39	4	94158.71	376634.85	39	281.48	-44.58	303.08	-48.00
40	1	82759.37	82759.37	40	284.99	0.00	287.68	0.00

Table 3.2
The Elaborations of Radial Quadrature for White Mere

These data contribute to the quadratures that estimate the Area of White Mere, and those findings are summarised in Table 3.3:-

Name of Lake or Surface:	White Mere	
Number of Radial Intervals	40	
Number of Perimeter Points	41	
Range (2π)	6.283185307179590	
$d\theta$	0.157079632679490	
Sine of $d\theta$	0.156434465040231	
θ_0	0.000000000000000	
θ_n	6.283185307179590	
Simpson Interval, h	0.157079632679490	
Mean Radius	280.468007986256000	
$\pi^*\mu_r^2$	247125	
$A_{Meister}$	248091	
A_{lakeOS}	255150	Meters^2
$A_{sectors}$	248091	Meters^2
$A_{Simpson}$	249792	Meters^2
PSD(A_{lakeOS},$A_{sectors}$)	2.766911	
PSD(A_{lakeOS},$A_{Simpson}$)	2.100150	
PSD(A_{lakeOS},$A_{Meister}$)	2.766911	
PSD(A_{lakeOS},$A_{\pi r^2}$)	3.145423	
$C_{lakeapprox}$	1971.042897	Meters
Radius of Reference Circle	284.985822	Meters
Area of Reference Circle	255150.476140	Meters^2
Circumference of Reference Circle	1790.618733	
Filename:	WhiteMereSmallHasRadialCropped.jpg	
Map:	NLS OS 25 inch 1892-1914 layer	
Nominal Center X:	966	Pixels
Nominal Center Y:	902	Pixels
Gravicenter X:	3.424170616	Meters
Gravicenter Y:	-3.613744076	Meters
Scale Unit:	100	Meters
Scale X left:	174	Pixels
Scale X right:	385	Pixels
Scale Length:	211	Pixels
Pixels per Meter:	2.11	Px/Meter

Table 3.3
Summary of Comparative Data for White Mere

Once again the Sectoral and Meister's Rule methods are seen to be numerically-equivalent but with this natural lake the Simpson's Rule areal estimate is little better than they with a PSD of 2.100150%.

This is undoubtedly due to the imprecision of the input data, both as regards the angular displacements and the radial readings. As I said, we need a young person with a steady hand and a sharp eye, and it does not help that my old PhotoDraw software works to only three-figure precision.

Figure 3.11 shows the re-constituted lake border in blue with its red reference circle. As you can judge, the two figures are of approximately equal area as they should be:-

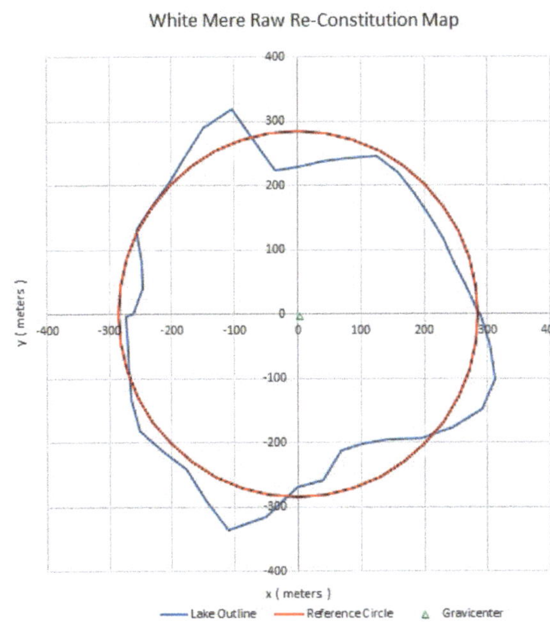

Figure 3.11
The Re-constituted Map of White Mere and its Reference Circle

Blake Mere

Blake Mere is a dumbbell-shaped eutrophic kettle hole centered about 52°53'58"N, 2°52'11"W.

Blake is an Old English word for "black" which probably reflects its historical color influenced by the peaty water of the district.

Blake is a common English surname, as is the Irish-influenced Moore, both alluding to a swarthy appearance.

Blake Mere seems to have been extensively altered by the works of The Shropshire Union (Llangollen) Canal at its southern shore. The Shropshire Union was cut hereabouts in 1834AD, a date former to the Ordnance Survey mappings of this lake. The local ground was, and is, unstable and there were significant engineering problems at Blake Mere in May 1834, but they were resolved and the Ellesmere stretch of the SUC opened for business in 1835AD.

Blake Mere's area was assessed by the Ordnance Survey to be 20.885 acres and this may have been taken at a Liverpool datum at about 312 feet during the Late Nineteenth-Century, or a modern UK Centre for Ecology and Hydrology UK Lakes Portal of 298.556 value in feet, probably against the Newlyn datum.

The maximum depth of this lake is about 13.5 meters and the mean depth 5.3. The UKCEH Catchment Based Data Hub ("authoritative") at https://data.catchmentbasedapproach.org/datasets/theriverstrust::uk-lakes-lakes-portal-ukceh/explore?location=52.903027%2C-2.864345%2C15 cites 5.93 hectares as the lake area. I take the OS nineteenth-century determination (in acres) as fiducial, and that is equivalent to 84519 square meters.

The Water Body UKCEH ID is 35052.

Blake Mere is a noted carp fishery and wading, swimming and boating are all prohibited. It is possible for the public to circumperambulate the lake on footpaths.

The small, round kettle lake called Kettle Mere drains into Blake Mere, but the latter has no clear subaerial outlet. I surmise that Blake Mere drains to the water table as the SUC is at a higher elevation.

Figure 3.12 shows Blake Mere and is taken from the angling site https://www.totalcarpmagazine.com/blakemere-shropshire.

The ecological state of Blake Mere is not known to me.

I constructed a 80-polar rose in red and superposed it on a 1901AD 25-inch-to-one-mile (1: 2534.4) OS map from the National Library of Scotland map archive at https://maps.nls.uk/geo/explore/side-by-side/#zoom=16.6&lat=52.89114 &lon=-2.86933&layers=168&right=ESRIWorld.

The resulting quadrature diagram is reproduced as Figure 3.13

I processed this diagram using the online utility PlotDigitizer, moving my mouse cursor by hand and recording the co-ordinates to fifteen decimal places by mouse left-click. I commenced at the Easternmost intersection of the red ray with the shore and worked sinistrally recording the co-ordinates as described until I had all eighty plus one (x,y) readings. I then took care to record the generated list to a .CSV file, and post-processed the records with the aid of EXCEL®.

After refinement, in particular the elimination of repeated co-ordinates, seventy-two useable data points "survived" for analysis.

This clarified data is reproduced in Table 3.4

Photo: James R Warren
(Olympus TG5: 10:44GMT 8 April 2026)

Figure 3.12
Blake Mere

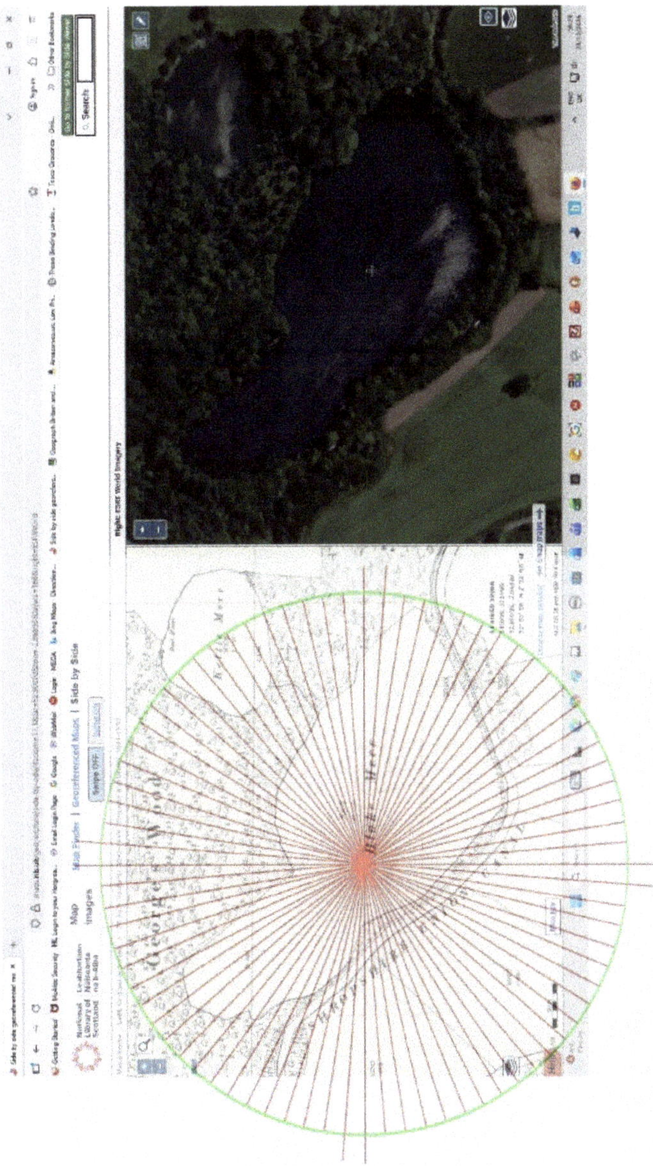

Figure 3.13
The Blake Mere Polar Quadrature Diagram

Subsequent to quadrature elaborations mediated by EXCEL®, the Input Data and Comparative Outcome Results were as summarised in Table 3.5

Once again, the numerical identity of the Sectoral and Meister quadrature methods is demonstrated, but paradoxically, though the interval density is almost double what it was for White Mere, the PSDs are even worse at around -4% in each case, the Simpson result worse than for the Sectoral.

This shows that finer sampling does not necessarily improve accuracy, especially where human judgement and frailties are involved.

Figure 3.14 shows the re-constituted lake border in blue with its red reference circle. As you can judge, the two figures are of approximately equal area as they should be.

Serial	Digitised Boundary Points in Ground Meters		Serial	Digitised Boundary Points in Ground Meters	
	x	y		x	y
0	545.016313	307.829906	37	184.668423	338.67617
1	510.321968	323.648493	38	195.707546	326.812222
2	499.282903	337.88523	39	237.498428	311.784546
3	489.820814	350.540118	40	241.440955	301.502477
4	474.050647	361.613126	41	247.749034	291.22035
5	474.050647	361.613126	42	250.903044	281.729192
6	459.857514	369.522434	43	257.211122	273.819913
7	444.875893	379.804503	44	259.576645	265.119665
8	435.413804	386.922871	45	261.942167	260.374086
9	422.797647	394.04124	46	266.673211	250.092017
10	413.335558	399.577758	47	272.981261	245.346438
11	402.296493	403.532398	48	277.712305	235.06437
12	395.199927	408.277977	49	283.231867	228.736912
13	383.372316	412.232646	50	290.328433	220.036664
14	374.698715	412.232646	51	299.002005	214.500174
15	362.871104	413.023556	52	309.252611	203.427137
16	351.831981	413.814496	53	318.714699	193.935979
17	342.369921	408.277977	54	328.176788	185.235789
18	333.696321	406.696127	55	340.004399	174.953663
19	322.657227	403.532398	56	354.197503	162.298805
20	315.56066	399.577758	57	375.487203	145.689278
21	306.887088	395.623119	58	387.314814	147.271157
22	295.847966	392.45939	59	405.450503	149.643946
23	283.231867	418.560075	60	418.066603	154.389525
24	268.250216	432.796812	61	434.625258	162.298805
25	248.537522	453.360978	62	448.818391	167.044384
26	204.381117	511.098867	63	466.165592	174.953663
27	179.14889	519.799086	64	481.147214	181.281121
28	158.647679	515.053506	65	501.648425	188.399489
29	143.666057	500.816769	66	522.149579	197.890648
30	134.203969	481.043542	67	545.016313	209.754595
31	136.569491	454.151918	68	567.094501	223.991333
32	132.626935	435.960541	69	595.480767	239.01898
33	149.18559	404.323338	70	588.3842	262.746876
34	155.493668	384.550082	71	567.094501	288.05665
35	159.436196	367.940555	72	545.016313	307.829906
36	172.840812	351.331028			

Table 3.4
The 72 Refined Cartesian Co-Ordinate Data for Blake Mere

Name of Lake or Surface:	Blake Mere	
Number of Radial Intervals	72	
Number of Perimeter Points	73	
Range (2π)	6.283185307179590	
$d\theta$	0.087266462599717	
Sine of $d\theta$	0.087155742747658	
θ_0	0.000000000000000	
θ_n	6.283185307179590	
Simpson Interval, h	0.087266462599717	
Mean Radius	156.269063569127000	
$\pi^*\mu_r^2$	76718	
$A_{Meister}$	87693	
A_{lakeOS}	84519	Meters^2
$A_{sectors}$	87693	Meters^2
$A_{Simpson}$	88033	Meters^2
$PSD(A_{lakeOS}, A_{sectors})$	-3.756126	
$PSD(A_{lakeOS}, A_{Simpson})$	-4.158348	
$PSD(A_{lakeOS}, A_{Meister})$	-3.756126	
$PSD(A_{lakeOS}, A_{\pi r}^2)$	9.229813	
$C_{lakeapprox}$	1343.205345	Meters
Radius of Reference Circle	164.021732	Meters
Area of Reference Circle	84518.671100	Meters^2
Circumference of Reference Circle	1030.578937	
Filename:	BlakeMereAndKettleMereScaledWith80RoseD.jpg	
Map:	NLS OS 25 inch 1892-1914 layer	
Nominal Center X:	308.4640936	Meters
Nominal Center Y:	307.5135243	Meters
Gravicenter X:	37.36648915	Meters
Gravicenter Y:	10.96536163	Meters
Scale Unit:	50	Meters
Scale X left:	67.7	Meters
Scale X right:	88.4	Meters
Scale Length:	20.7	Meters

Table 3.5
Summary of Comparative Data for Blake Mere

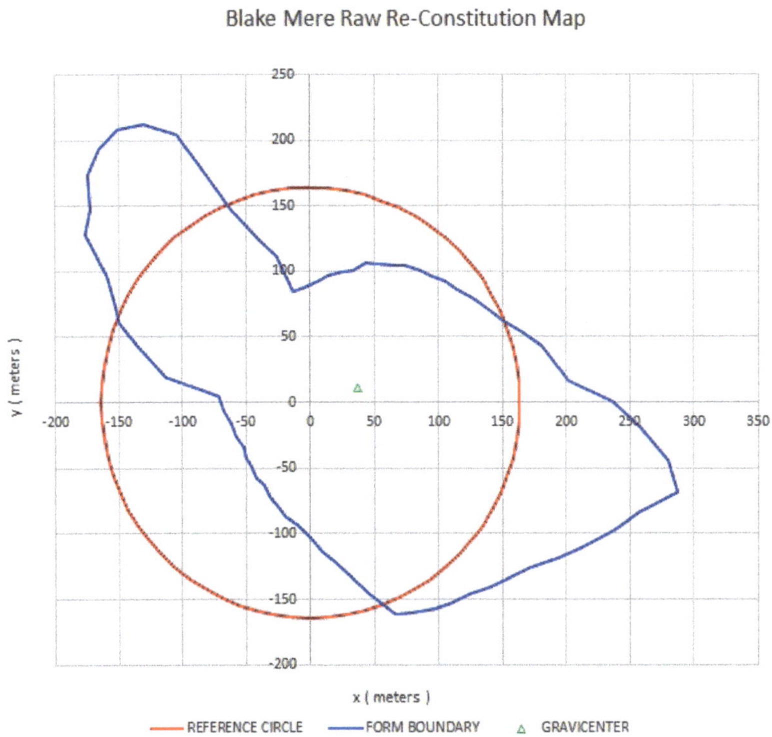

Figure 3.14
The Re-constituted Map of
Blake Mere
And its Reference Circle

<u>Ellesmere</u>

Ellesmere the lake is immediately East of Ellesmere Town and the lake is centered at 52°54'31" N, 2°53'3" W.

Ellesmere Town is named for Ellesmere the lake, the name of that latter deriving from the European eel (Anguilla anguilla), a catadromous edible snake-like fish with a tough, slimy skin which formerly was tanned for leather. Eels were very common as adult fish in English lakes and were a prized food source.

According to AI Overview the name derives from the Old English "Elli's Lake" which means the lake of the noble. I am a republican and I prefer to award eels the estate! God gave the lake to the eels, and they make the best use of it.

Although Ellesmere is the largest of the Salopian Lake District kettles it does not appear to feature upon the DEFRA water bodies database, nor the UK CEH Catchment Data Hub. Accordingly, I can offer you no official British Government Water Body ID.

Notwithstanding that, Google AI Overview declares that the Surface Area of Ellesmere varies from 43 to 46 hectares whilst its Maximum Depth is 18.8 meters and its Mean Depth 7.6 meters. If the average surface area is 44.5 hectares then the Indicated Volume is 3.382 million cubic meters and the nominal contained mass 3.38 million tonnes.

I am not au fait with the lake's ecological quality accept that it is said to have "moderate" nitrogen and phosphorous levels and a relatively high salinity. Carp, pike, perch and bream all live in Ellesmere today, and grey heron and other water fowl are abundant. The muddy bottom supports invertebrates luxuriantly. Like many Shropshire lakes Ellesmere has a rich and Protected diversity of marginal vegetation, reeds and water-lilies and other species.

After obtaining a permit you may use your own sail or rowing boat upon Ellesmere, a singularly pleasant experience, but fishing is only by rod and a Licence is difficult to obtain. Swimming is prohibited and unsafe.

Ellesmere's area was assessed by the Ordnance Survey to be 113.85 acres. On 21 February 1901 the surface of the lake was found to be 292.2 feet and this probably relates to the tidal high-water datum at Liverpool.

I constructed a 80-polar rose in red and superposed it on a 1901AD 25-inch-to-one-mile (1: 2534.4) OS map from the National Library of Scotland map archive at https://https://maps.nls.uk/geo/explore/side-by-side/#zoom=17.6&lat=52.90861&lon=-2.88532&layers=168&right=ESRIWorld

Figure 3.15 is a photograph of Ellesmere and Figure 3.16 is my quadrature diagram.

Figure 3.17 is the Re-constituted map of Ellesmere and its Reference Circle.

I processed the quadrature diagram using the online utility PlotDigitizer, moving my mouse cursor by hand and recording the co-ordinates to fifteen decimal places by mouse left-click. I commenced at the Easternmost intersection of the red ray with the shore and worked sinistrally recording the co-ordinates as described until I had all eighty plus one (x,y) readings. I then took care to record the generated list to a .CSV file, and post-processed the records with the aid of EXCEL®.

After refinement, in particular the elimination of repeated co-ordinates, eighty-one useable data points "survived" for analysis.

This clarified data is reproduced in Table 3.6

Subsequent to quadrature elaborations mediated by EXCEL®, the Input Data and Comparative Outcome Results were as summarised in Table 3.7

On this occasion the Sectoral and Meister area estimates are *not* numerically equivalent.

Photo: James R Warren
(Olympus TG5: 09:09GMT 8 April 2026)

Figure 3.15
Ellesmere

Reference to Table 3.7 shows that once again the PSD accuracy is near to ±2.6% whether Sectoral, Simpson or Meister's Rule is used, though in this case Sectoral area estimation is marginally better than Simpsons Rule, whilst Meister's Rule is rather better than either.

The negativity of all the PSD numbers indicates that, contrary to theoretic possibilities, each of the three methods *over*estimates the area relative to the fiducial.

In my assessment, these surprising if small discrepancies do not belie the principles, but result from the uneven intervals and poor quality of the data, as gathered by the eyes and trembling hands of a seventy-three year old.

As they say, "Garbage in: Garbage out".

I am sure that if a machine could be trained to read co-ordinates at the intersections of lake shores with polar rays, with perhaps 2048 radial intervals, then these discrepancies would largely vanish.

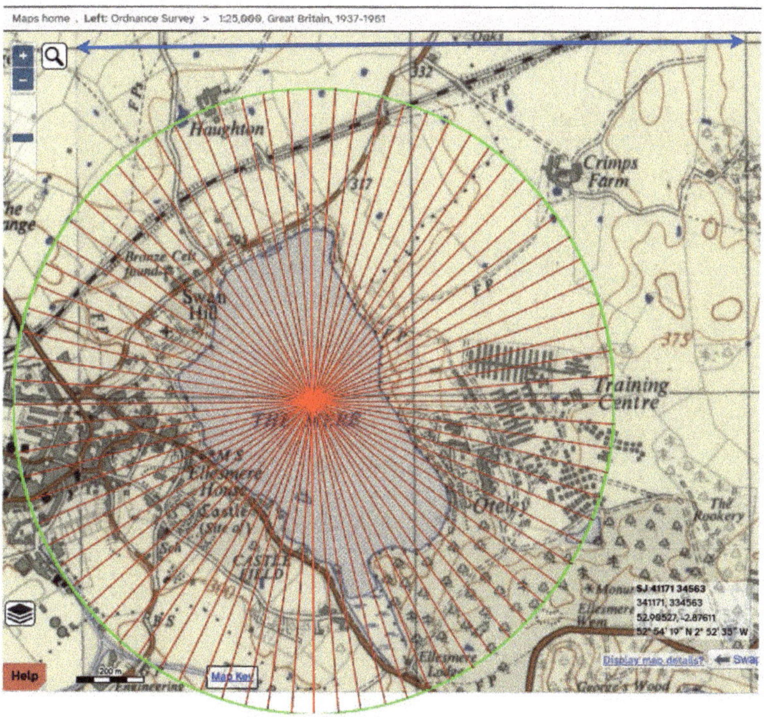

Figure 3.16
The Ellesmere Polar Quadrature Diagram

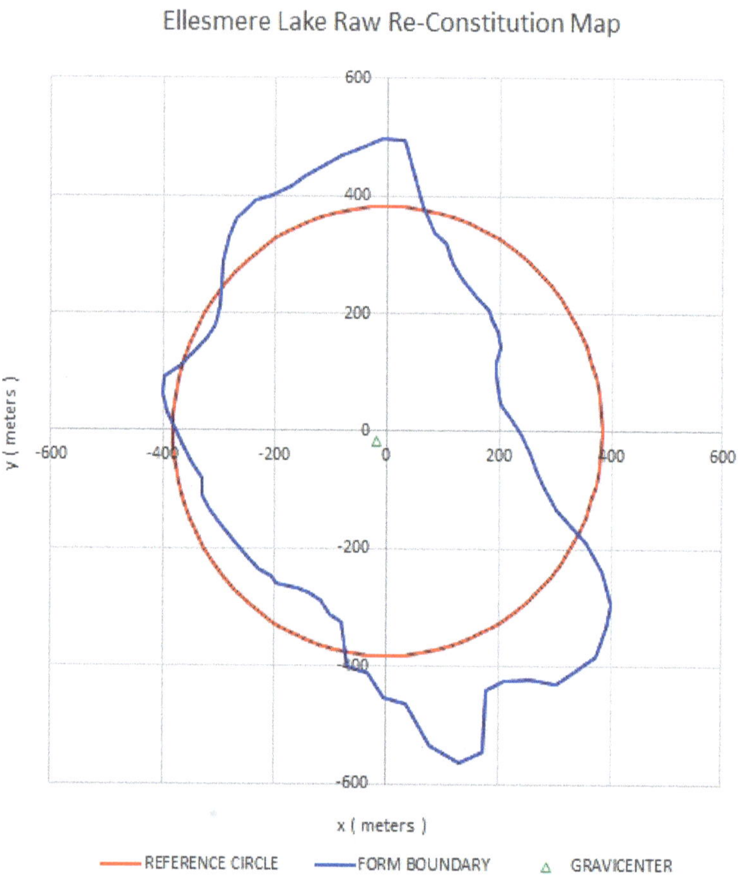

Figure 3.17
The Re-constituted Map of Ellesmere and its Reference Circle

Serial	Digitised Boundary Points in Ground Meters		Serial	Digitised Boundary Points in Ground Meters	
	x	y		x	y
0	1143.3517	894.935609	41	542.521648	869.726155
1	1128.14081	917.624118	42	557.732535	844.516702
2	1120.53537	927.707899	43	575.47857	819.307248
3	1107.85963	947.875462	44	575.47857	789.055903
4	1102.78933	975.605861	45	588.15431	766.367395
5	1100.25418	995.773424	46	605.900345	743.678887
6	1100.25418	1018.46193	47	623.64638	723.511324
7	1107.85963	1043.67139	48	636.32212	708.385651
8	1102.78933	1071.40179	49	659.138451	683.176197
9	1092.64874	1091.56935	50	676.884486	665.52958
10	1085.0433	1106.69502	51	699.700817	652.924853
11	1064.76211	1126.86258	52	709.841408	640.320126
12	1052.08637	1144.5092	53	745.333479	632.75729
13	1036.87549	1162.15582	54	768.14981	625.194454
14	1021.6646	1187.36527	55	788.430993	612.589727
15	1008.98886	1220.13756	56	803.64188	587.380273
16	988.707676	1237.78418	57	826.458211	574.775546
17	968.426493	1285.68214	58	836.598803	499.147185
18	935.46957	1394.08279	59	872.090873	489.063403
19	897.442352	1396.60374	60	902.512648	446.207332
20	864.48543	1383.99901	61	940.539866	436.12355
21	823.923063	1368.87334	62	983.63738	365.53708
22	798.571585	1356.26861	63	1036.87549	337.806681
23	760.544366	1333.58011	64	1077.43785	355.453298
24	732.657739	1315.93349	65	1082.50815	461.333004
25	699.700817	1300.80782	66	1115.46507	476.458676
26	669.279042	1293.24498	67	1115.46507	476.458676
27	636.32212	1260.47269	68	1163.63288	478.979622
28	623.64638	1230.22134	69	1206.73039	471.416786
29	610.970641	1187.36527	70	1244.75761	494.105294
30	608.435493	1147.03015	71	1277.71454	516.793803
31	608.435493	1147.03015	72	1297.99572	567.21271
32	605.900345	1111.73691	73	1305.60116	607.547836
33	598.294901	1081.48557	74	1287.85513	660.487689
34	583.084014	1056.27611	75	1259.9685	710.906597
35	565.337979	1038.6295	76	1206.73039	766.367395
36	534.916204	1010.8991	77	1186.44921	804.181576
37	507.029577	990.731534	78	1173.77347	829.391029
38	504.49443	963.001134	79	1161.09773	862.163319
39	512.099873	930.228845	80	1143.3517	894.935609
40	527.310761	899.9775			

Table 3.6
The Eighty-One Refined Cartesian Co-ordinate Data for Ellesmere

Name of Lake or Surface:	Ellesmere	
Number of Radial Intervals	80	
Number of Perimeter Points	81	
Range (2π)	6.283185307179590	
$d\theta$	0.078539816339745	
Sine of $d\theta$	0.078459095727845	
θ_0	0.000000000000000	
θ_n	6.283185307179590	
Simpson Interval, h	0.078539816339745	
Mean Radius	375.277462425238000	
$\pi^*\mu_r^2$	442440	
$A_{Meister}$	471510	
A_{lakeOS}	460735	Meters^2
$A_{sectors}$	472066	Meters^2
$A_{Simpson}$	472217	Meters^2
PSD(A_{lakeOS},$A_{sectors}$)	-2.459348	
PSD(A_{lakeOS},$A_{Simpson}$)	-2.492025	
A_{lakeOS} (Less Moscow Island)	459683	Meters^2
$A_{sectors}$	472066	Meters^2
$A_{Simpson}$	472217	Meters^2
PSD(A_{lakeOS},$A_{sectors}$)	-2.693830	
PSD(A_{lakeOS},$A_{Simpson}$)	-2.726581	
PSD(A_{lakeOS},$A_{Meister}$)	-2.572775	
PSD(A_{lakeOS},$A_{\pi r^2}$)	3.750965	
$C_{lakeapprox}$	2948.393399	Meters
Radius of Reference Circle	382.957581	Meters
Area of Reference Circle	460735.011000	Meters^2
Circumference of Reference Circle	2406.193447	
Filename:	EllesmereHasGridHasRose80aScreenshot 2026-01-05 095501.gif	
Map:	OS 1:25000, Great Britain, 1937-1961	
Nominal Center X:	905.0477958	Meters
Nominal Center Y:	899.9775	Meters
Gravicenter X: (Raw Meters)	897.5374202	Meters
Gravicenter Y: (Raw Meters)	893.769672	Meters
Gravicenter X:	-18.82347306	Meters
Gravicenter Y:	-17.45754674	Meters
Scale Unit:	50	Meters
Scale X left:	67.7	Meters
Scale X right:	88.4	Meters
Scale Length:	20.7	Meters

Table 3.7
Summary of Comparative Data for Ellesmere

PART FOUR
SOME IDEAS FOR EXPLORATION

An idea is a precious thing but most ideas are erroneous in conception, or unprofitable in practice.

An idea is a child of the mind and as a child it demands nurture and fairness, but not indulgence, or, God forbid, idolisation. Any ethical idea is worthy of trial and we will try it if we are true scientists, humble under God.

Most of my ideas are nonsensical: The art is to know *which* ideas.

Do not be afraid to add your ideas to the Great Debate, as the heroes of the past, Christian and Pagan, did in the name of Progress. We are their heirs, but usually prove unworthy.

We must eschew interpretations of interpretations, remembering always that inference breeds error, and errors compound with each revision.

Always refer back to The Field where wills and forces play in their native state, but bear in mind that our exploration is only as good as our senses, and that though mathematics is a work of man we approach via Algebra or via Geometry.

The Fitting and Integration of a Two-Interval Parabola

You can see that a number of desiderata are missing from the methods that we have so far explored.

In particular, we have glossed over the desirability of using equal abscissal intervals (x-intervals) when using Simpson's Rule. Equal intervals were easy enough to achieve when we computed "exact" quadratures of formal mathematical equations as derived by classical geometers and scientists.

But data that arise from the natural environment, such as lake shores or cloud profiles or tree sections are seldom conveniently a set distance apart, even after our best efforts competently to survey.

We are entitled to suspect that the failure to input data at precisely equivalent intervals has contributed to the imprecision of our area estimation outcomes, along with warping of paper charts over their years of storage, frailty of hand-and-eye, and failures to assume proper sighting stations in the field, as well as other vagaries too numerous to list.

Without unprofitable excursions into advanced mathematical theory, which may or may not enhance our results, we can investigate several arithmetic tricks which may ameliorate if not surmount our current problems.

Imagine that we have two adjacent intervals, one of which is wider than the other, but we have the necessary and sufficient three measured Cartesian co-ordinates that will permit us to calculate the unique parabola that connects them.

The situation is illustrated by the sketch plot Figure 4.1:-

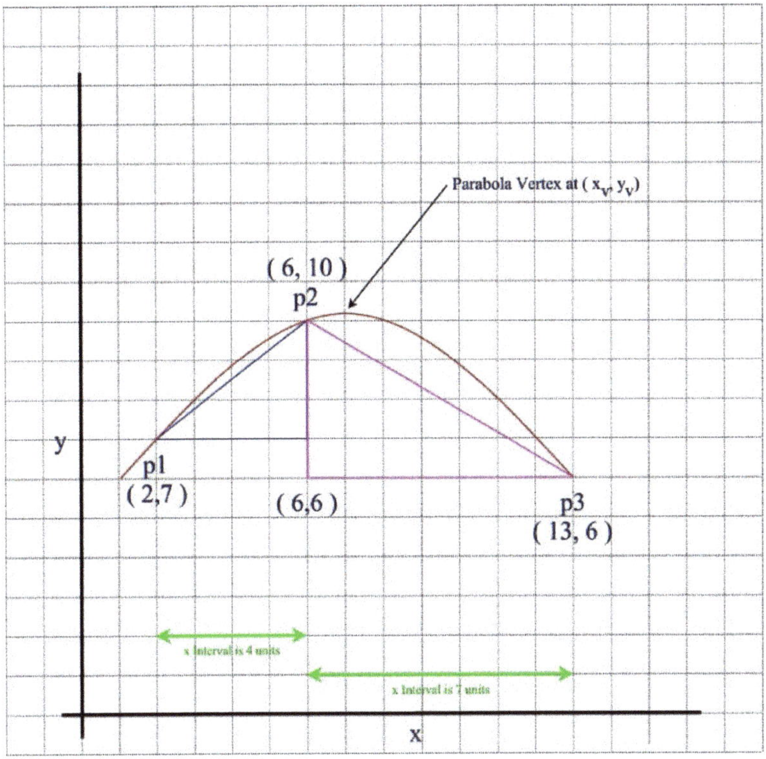

Figure 4.1
Fitting a Sketch Curve to Three Points

The arrowed green bars show clearly that the left-hand interval is four units wide, and the right-hand seven.

The red line is supposed to be a parabola, i.e. a second-degree algebraic curve, connecting point p1, p2 and p3, *which are the only external data present.*

These data have the following Cartesian co-ordinates:-

p1 $(xp_1, yp_1) = (2,7)$
p2 $(xp_2 \ yp_2) = (6,10)$
p3 $(xp_3, yp_3) = (13,6)$

The parabolic vertex (x_v, y_v) *appears* to be at (7, 10.2). This is not precisely correct. The reason for the discrepancy is that I used PhotoDraw to fit a "curve". I suspect that PhotoDraw's "curve" is in fact a cubic spline, i.e. a third-degree polynomial, subtly different to a parabola.

Equation 4.1 is a cubic equation:-

$$y = a.x^3 + b.x^2 + c.x^1 + d$$

Equation 4.1
The Cubic Equation

which is similar to but differs from a quadratic equation:-

$$y = a.x^2 + b.x + c$$

Equation 4.2
The Quadratic Equation for a Parabola

If we are to approximate area in an interval using a parabola in the style of Simpson's Rule but maintaining accuracy despite variant interval sizes we do not need to invoke matrices, Gauss-Jordan Elimination or any advanced technique whatsoever.

We just need to obtain values for the Term Constants a, b and c and then apply the most simple analytic calculus to perform a truly basic analytic integration for area-in-an-interval. This will be done with no vitiation of numerical accuracy.

We even have a choice of equations for a, b and c: Leibovici's Formulae and the classical formulae as presented by Wolfram Alpha® and others.

Leibovici's Formulae

$$a = \frac{xp_1(yp_3 - yp_2) + xp_2(yp_1 - yp_3) + xp_3(yp_2 - yp_1)}{(xp_1 - xp_2)(xp_1 - xp_3)(xp_2 - xp_3)}$$

Equation 4.3

$$b = \frac{yp_2 - yp_1}{xp_2 - xp_1} - a \times (xp_1 + xp_2)$$

Equation 4.4

$$c = yp_1 - a.xp_1{}^2 - b.xp_1$$

Equation 4.5

Wolfram Alpha® Formulae

$$denom = (xp_1 - xp_3)(xp_2 - xp_3)(xp_1 - xp_2)$$

Equation 4.6

$$a = \frac{yp_1(xp_2 - xp_3) + yp_2(xp_3 - xp_1) + yp_3(xp_1 - xp_2)}{denom}$$

Equation 4.7

$$b = \frac{yp_1(xp_3{}^2 - xp_2{}^2) + yp_2(xp_1{}^2 - xp_3{}^2) + yp_3(xp_2{}^2 - xp_1{}^2)}{denom}$$

Equation 4.8

$$c = \frac{yp_1(xp_2{}^2.xp_3 - xp_3{}^2.xp_2) + yp_2(xp_3{}^2.xp_1 - xp_1{}^2.xp_3) + yp_3(xp_1{}^2.xp_2 - xp_2{}^2.xp_1)}{denom}$$

Equation 4.9

If there is a criticism to be made of the Leibovici formulation it is that the recycling of prior-computed term constants may introduce error.

But in practice both the Leibovici and the Wolfram equations produce values of a, b and c that are identical and accurate to fifteen figures using MathCad®.

So your choice will hinge upon which sequence of calculations is the most cheap in terms of time and resources.

The computed values of the quadratic equation constants are:-

denom	-308
a	-0.12012987012987
b	1.71103896103896
c	4.05844155844156

Elaboration

To help plot our parabola as a nice red line it is convenient to elaborate our polynomial equation at suitably close equal intervals of x.

I have chosen to define n = twenty-two intervals of x_i, for $x_i = x_0 \ldots x_n$, giving twenty three values of paired trial data (x_i, y_i).

Furthermore, I wish to Integrate this range (i.e. find the area under the curve) between the Lower Bound lb = 2, and the Upper Bound ub = 13 so that the equal common Abscissal Interval, h, is:-

$$h = \frac{ub - lb}{n} = 0.5$$

Equation 4.10

and:-

$$x_i = lb + i.h$$

Equation 4.11a

$$y_i = a.x_i{}^2 + b.x_i + c$$

Equation 4.11b

The co-ordinates of the parabolas vertex, its apex, are in fact:-

$$x_v = \frac{-b}{2a} = 7.12162162162162$$

Equation 4.12a

$$y_v = c - \frac{b^2}{4a} = 10.1511275886276$$

Equation 4.12b

These features are more accurately represented in the EXCEL®-drafted plot diagram Figure 4.2:-

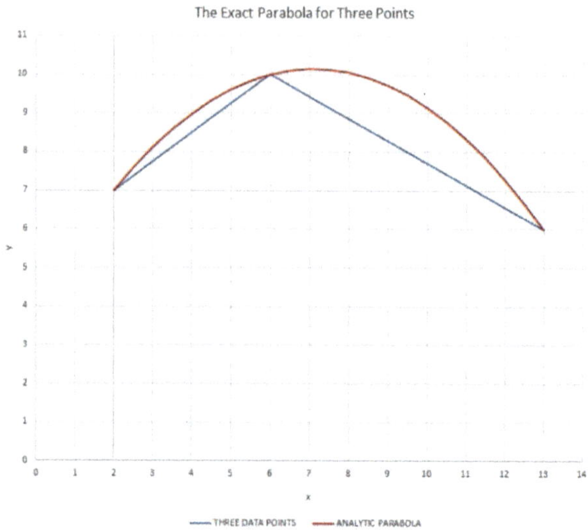

Figure 4.2
Parabola and its Chords
Resolved Diagram

The inputs and results are summarised in Table 4.1

The Analytic Integral

The general analytic integral of an algebraic exponent is:-

$$\int x^n . dx = \frac{1}{n+1} x^{n+1} + C$$

Equation 4.13

where n is any Integer and C is a Constant of Integration.

Equation 4.13 may be expanded for a sum of the terms in a second-degree algebraic polynomial (i.e. a parabola equation) as:-

$$\int (a.x^2 + b.x + c).dx = a.\frac{x^3}{3} + b.\frac{x^2}{2} + c.x + C$$

Equation 4.14

Of particular interest to us is the finite area under the parabola defined as the Definite Integral bounded in x by lb and ub.

This can be defined as:-

$$\int_{lb}^{ub} (a.x^2 + b.x + c).dx = \left[a.\frac{x^3}{3} + b.\frac{x^2}{2} + c.x\right]_{lb}^{ub}$$
$$= \left(a.\frac{ub^3}{3} + b.\frac{ub^2}{2} + c.ub\right) - \left(a.\frac{lb^3}{3} + b.\frac{lb^2}{2} + c.lb\right)$$
$$= 98.1488095238095\ \textit{square units}$$

Equation 4.15

Note that the Constant of Integration, C, drops out of the reconning because it is essentially subtracted from itself.

Number of Intervals	22
Lower Limit of Integration (x_1)	2
Upper Limit of Integration (x_2)	13
Equal Interval Width	0.5

Top of Definite Integral	109.367424
Base of Definite Intergal	11.2186147
Definite Integral	98.1488095

	x	**y**	**Intermediate Differences**			**Analytic Constants**	**Claude Leibovici**	**Wolfram Alpha**
						denom		-308
P_1	2	7	X1	(x1-x2)	-4	**a**	-0.12012987	-0.12012987
P_2	6	10	X2	(x1-x3)	-11	**b**	1.711038961	1.711038961
P_3	13	6	X3	(x2-x3)	-7	**c**	4.058441558	4.058441558
			Y1	(y3-y2)	-4			
			Y2	(y1-y3)	1			
			Y3	(y2-y1)	3			

P_v	7.122	10.15
P_v	7.122	10.15

Serial	**x**	**CL**	**WA**
0	2	7	7
1	2.5	7.585227	7.585227
2	3	8.11039	8.11039
3	3.5	8.575487	8.575487
4	4	8.980519	8.980519
5	4.5	9.325487	9.325487
6	5	9.61039	9.61039
7	5.5	9.835227	9.835227
8	6	10	10
9	6.5	10.10471	10.10471
10	7	10.14935	10.14935
11	7.5	10.13393	10.13393
12	8	10.05844	10.05844
13	8.5	9.92289	9.92289
14	9	9.727273	9.727273
15	9.5	9.471591	9.471591
16	10	9.155844	9.155844
17	10.5	8.780032	8.780032
18	11	8.344156	8.344156
19	11.5	7.848214	7.848214
20	12	7.292208	7.292208
21	12.5	6.676136	6.676136
22	13	6	6

Table 4.1
Parabola Equation and Fitting

The Primitive Simpson Integral

The Primitive Simpson Integral, I_{SPrim}, is defined by:-

$$I_{SPrim} = \frac{1}{3} \cdot \frac{ub - lb}{2} \cdot \left(y_0 + 4. y_{\frac{ub-lb}{2}} + y_n \right)$$

Equation 4.16

or for simplicity:-

$$I_{SPrim} = \frac{ub - lb}{6} \cdot (y_0 + 4. y_{Mid} + y_n)$$

Equation 4.17

This is the most basic form of Simpson's Rule: The so-called "Kepler Barrel Rule", and is exact for any parabola.

For our context we may solve this as:-

$$I_{SPrim} = \frac{1}{3} \cdot \frac{13 - 2}{2} \cdot (y_0 + 4. y_{11} + y_{22})$$
$$= 98.1488095238095$$

Equation 4.18

Because we are integrating a parabola, for which Simpson's Rule is perfect, we have exactly the same outcome number as for the analytic Definite Integral.

The Extended Simpson's Rule

We have seen that Simpson's Rule can be extended to any even number of intervals. Simpson's Rule essentially treats each interval curve segment as the arc of a parabola, even when it clearly is not. Notwithstanding, as long as we have *sufficient* intervals, Simpson's Rule will yield a useful approximation of the area under the curve in the majority of cases.

As an extended Newton-Coates series, Extended Simpson's Rule requires an Alternator for the interior terms, which Alternator we may define as:-

$$ALT(i) = 3 + (-1)^{i+1}$$

Equation 4.19

So that the Extended Simpson's Rule Integral, I_{SExt}, in Cartesian form, is:-

$$I_{SExt} = \frac{h}{3} \cdot \left[y_0 + \sum_{i=1}^{n-1} (ALT(i) \times y_i) + y_n \right]$$

Equation 4.20

I_{SExt} also computes to 98.1488095238095, because all arcs of a parabola are themselves parabolas.

So in principle you can use elementary calculus to compute the areas of survey intervals *whatever their individual widths*. You can use any even number of intervals and effectively calculate their areas two at a time to get a reasonable estimate of total area, so if you have divided your lake into a raster of 41 parallel ordinates, you can, in theory, apply the calculus to only 20 iterations involving 20 pairs of subareas.

Things then devolve to questions of affordability and resources.

Lagrange Interpolation

Lagrange Interpolation artificially fits y-values equally-spaced in x to an arbitrary set of input (x, y) readings from the environment.

I must stress from the outset: This does not magically create bonus data for us to work with. It just assists our management of other, actually effective, methods.

If anything, Lagrange Interpolation slightly spoils output information through arithmetic error.

This is not to deny the utility of Lagrange Interpolation when judiciously applied.

The general equation is:-

$$P(x) = \sum_{i=0}^{n} y_i \cdot L_i(x)$$

Equation 4.21

where $L_i(x)$ is the Lagrange Basis Polynomial defined by:-

$$L_i(x) = \prod_{j=0, j\neq i}^{n} \frac{x - x_j}{x_i - x_j}$$

Equation 4.22

Excluding the singularity defined by j = i can be tricky in some elaborative systems, but you can always try to split the product series into two parts, perhaps:-

$$L_i(x) = \prod_{j=0}^{j-1} \frac{x - x_j}{x_i - x_j} \times \prod_{j=j+1}^{n} \frac{x - x_j}{x_i - x_j}$$

Equation 4.23

Fortunately, when there are only three input data points; (xp_1, yp_1), (xp_2, yp_2),(xp_3, yp_3), we can work with:-

$$P_i(x) = yp_1 \cdot \frac{(x_i - xp_2)(x_i - xp_3)}{(xp_1 - xp_2)(xp_1 - xp_3)} + yp_2 \cdot \frac{(x_i - xp_1)(x_i - xp_3)}{(xp_2 - xp_1)(xp_2 - xp_3)} + yp_3 \cdot \frac{(x_i - xp_1)(x_i - xp_2)}{(xp_3 - xp_1)(xp_3 - xp_2)}$$

Equation 4.24

Using Equation 4.24 we may tabulate the interpolated values as in Table 4.2

Note the exact coincidence at the three input data points and the evenness of the gaps between those and the nine interpolated points. I have chosen the x-spacing to be unity, but you could use any arbitrary interval as long as it formed at least three equally-spaced ordinates.

The interpolated points are shown as green squares in Figure 4.3

It is in the nature of Lagrange Interpolation that it only yields plot points, not the polynomial equation, and in particular, not the Term Coefficients that define a Polynomial Equation.

LAGRANGE INTERPOLATION		
i	x_i	$P_i(x)$
0	2	7
1	3	8.11038961038961
2	4	8.98051948051948
3	5	9.61038961038961
4	6	10
5	7	10.1493506493506
6	8	10.0584415584416
7	9	9.72727272727727
8	10	9.15584415584442
9	11	8.34415584415558
10	12	7.29220779220078
11	13	6

Table 4.2
Interpolated Lagrange Points

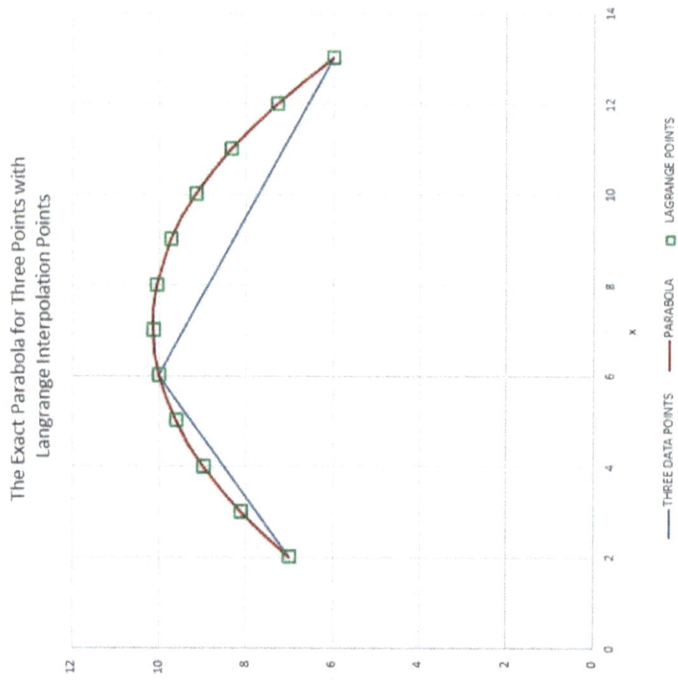

Figure 4.3
The Interpolated Lagrange Points on their Parabola

The Archimedes Parabola Theorem (APT)

Refer to Figure 4.4 which reproduces the part of Figure 4.2 focused upon a dextral interval between x = 6 and x = 13.

The essence of Archimedes Parabola Theorem (APT) is this:-

Line AC is a chord of the parabola in the interval {x: 6,13 }

Line UV is a tangent to the parabola and is parallel to AC.

Therefore AC and UV have the same gradient, and if the gradient of AC, dy/dx_{AC}, can be determined, then that is the gradient of the equation of the parabola where the tangent just touches the parabola. This point we shall call (x_B , y_B).

Draw the triangle ABC between (x_1, y_1), (x_B, y_B) and (x_2, y_2).

The Area of the Parabolic Segment ABC (that has the base AC) is 4/3 times the Area of that Triangle ABC.

Clearly, if we can compute and add the Area of the trapezium ACDE we can infer the Area of the Whole Interval from straight-line geometry only, without recourse to even elementary calculus.

Please note that contrary to much of the discussion of APT on the Internet, we do ***not*** require knowledge of any altitude of the triangle.

What we do need is precise knowledge of the position of the triangle apical Point B which has the Cartesian co-ordinates (x_B, y_B).

Notwithstanding, we shall in due course calculate the altitude BF!

Also, in this context do note that the triangle ABC area is ***not*** AC.BF/2

The sloping lines of the Figure 4.4 system are all hypotenuses and therefore it is computationally convenient to define a function PYTHAG for The Pythagorean Theorem at the outset:-

$$PYTHAG(op, ad) = \sqrt{op^2 + ad^2}$$

Equation 4.25
The Function PYTHAG

Where PYTHAG(op,ad) is the Length of the Hypotenuse to be found; op is the Pythagorean Opposite and ad is the Pythagorean Adjacent.

Table 4.3 lists the key lineament values *as determined from the diagram.*

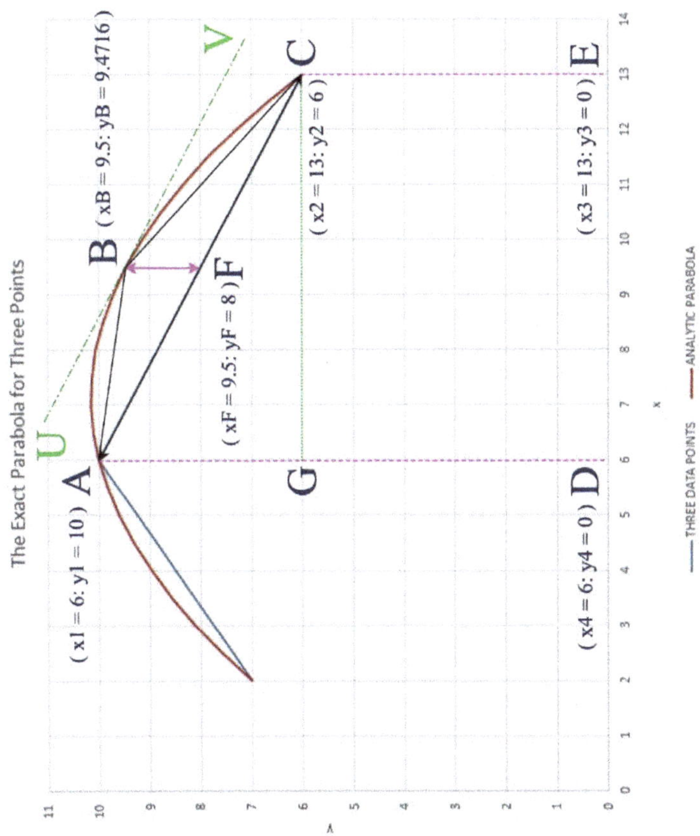

Figure 4.4
The Scheme of the Area of an Arbitrary Cartesian Parabolic Interval by applying Archimedes Parabola Theorem

Line	Adjacent	Opposite	Length in Native Units
AD			10
DE			7
EC			6
AC	4	7	8.06225775
AB	0.5	3.5	3.53553391
BC	3.5	3.5	4.94974747
BF			1.46020930
AF	3.5	2	4.03112887
FC	3.5	2	4.03112887
AF+FC			8.06225775

Table 4.3
Length of Principal Lines in The APT System

A further admonition I am afraid: Please do not assume that AF and FC are invariably equal: They only happen to be so in this example.

Having all sides of the triangle ABC to hand as AB = 3.53553391, BC = 4.94974747 and AC = 8.06225775 we reliably may calculate the Area of the Included Triangle ABC, A_{Heron}, as:-

$$s = \frac{AB + BC + AC}{2} = 8.27376956126856$$

Equation 4.26a

$$A_{Heron} = \sqrt{s(s - AB)(s - BC)(s - AC)} = 5.25000000000001$$

Equation 4.26b

Discounting the obvious error of computation the triangle area is exactly 5.25 units.

Accordingly, by the Archimedes Parabola Theorem, the Area of Segment ABC, A_{seg}, is:-

$$A_{seg} = \frac{4}{3} \cdot A_{Heron} = 7.00000000000001$$

Equation 4.27

Meanwhile the Area of Trapezium ADEC, A_{ADEC}, is given by:-

$$A_{ADEC} = DE.EC + \frac{DE.AG}{2} = 56$$

Equation 4.28

Therefore, the Area of the Entire Interval ABCED is:-

$$A_{interval} = A_{ADEC} + A_{seg} = 63$$

Equation 4.29

Corroboration by Analytic Integral Calculus

We know that the Lower Bound (in x) of the Interval, lb, is 6; and the Upper Bound, ub, is 13.

Further, from previous work we know that the Polynomial Co-efficients are a = -0.12012987012987, b = 1.71103896103896 and c = 4.05844155844156

By using the MathCad® definite integral function we establish that:-

$$I_{interval} = \int_{lb}^{ub} (a.x^2 + b.x + c).dx = 62.8674242424243$$

Equation 4.30

where $I_{interval}$ is the Area of Interval ABCED in square units.

To check the integrity of this integration we may follow through manually in this way:-

$$\begin{aligned} I_{interval} &= \int_{lb}^{ub} (a.x^2 + b.x + c).dx \\ &= \left[a.\frac{x^3}{3} + b.\frac{x^2}{2} + c.x\right]_{lb}^{ub} \\ &= \left[a.\frac{ub^3}{3} + b.\frac{ub^2}{2} + c.ub\right] - \left[a.\frac{lb^3}{3} + b.\frac{lb^2}{2} + c.lb\right] \\ &= 109.367424242424 - 46.5 \\ &= 62.8674242424243 \end{aligned}$$

Equation 4.31

Taking $A_{interval}$ as fiducial, the Percentage Specific Defect between $A_{interval}$ and $I_{interval}$ is:-

$$PSD(A_{interval}, I_{interval}) = 0.210437710437699$$

Equation 4.32

showing that the integration under-estimates the total area by 0.21%. This may be partially due to the fact that the height of the parabola at its summit is less than that suggested by the diagram. The rest may be computational error.

Corroboration by the Primitive Simpson's Rule

Primitive Simpson's Rule (i.e. Kepler's Barrel Rule) is exact for a parabola.

By way of predicate:-

$$y_{lb} = a.lb^2 + b.lb + c = 10$$

Equation 4.33a

$$y_{ub} = a.ub^2 + b.ub + c = 6.00000000000001$$

Equation 4.33b

and:-

$$x_{Mid} = lb + \frac{ub - lb}{2} = 9.5$$

Equation 4.34a

$$y_{Mid} = a.x_{Mid}{}^2 + b.x_{Mid} + c = 9.47159090909091$$

Equation 4.34b

We are able to set the interval h to unity as our system is calibrated in arbitrary units x = 0,1,14.

Hence the Area of the Interval by Simpson's Rule, I_{Sprim}, is:-

$$I_{Sprim} = \frac{h}{3} \cdot \frac{ub - lb}{2} \cdot (y_{lb} + 4y_{Mid} + y_{ub}) = 62.8674242424243$$

Equation 4.35

If the analytic $I_{interval}$ is fiducial then:-

$$PSD(I_{interval}, I_{Sprim}) = 0.000000000000011$$

Equation 4.36

I ascribe the tiny discrepancy to computational error.

The Analytic Determination of Triangle Altitude BF

The key to this solution, if the solution is required, is to note that the gradient of the parabola g_{para} is equal to the gradient g_{AC} of its chord AC at the touching tangent because the tangent line UV is parallel to AC.

In algebra:-

$$g_{para} = \frac{dy}{dx}_{para} \equiv g_{AC} = \frac{dy}{dx}_{AC}$$

Equation 4.37

The Gradient g_{para} of the parabola at x_B is:-

$$g_{para} = \frac{dy}{dx}_{para} = a.2.x_B + b = -0.57142857142857$$

Equation 4.38

By differentiation of the equation of the parabola.
The Gradient g_{AC} of the chord is:-

$$g_{AC} = \frac{dy}{dx}_{AC} = \frac{-AC}{DE} = -0.571428571428571$$

Equation 4.39

The PSD is -0.00000000000233 which I ascribe to computational error.

By transposition for x_B we obtain:-

$$x_B = \frac{\frac{-AG}{DE} - b}{2a} = \frac{\frac{-(y_1 - y_2)}{(x_3 - x_4)} - b}{2a} = \frac{1}{2a} \cdot \left[\frac{-(y_1 - y_2)}{(x_3 - x_4)} - b \right]$$
$$= 9.50000000000001$$

Equation 4.40

We may now obtain y_B by substitution of x_B in the equation of the parabola:-

$$y_B = a.x_B{}^2 + b.x_B + c = 9.47159090909091$$

Equation 4.41

Knowing by definition that the x-coordinate of the point F is the same as x_B, i.e. $x_F = x_B$, we may move forward to compute y_F as:-

$$\begin{aligned} y_F &= y_1 - \frac{-AG}{DE} \cdot (x_F - x_3) \\ &= y_1 - \frac{AG}{DE} \cdot (x_3 - x_F) \\ &= y_1 - \frac{(y_1 - y_2)}{(x_3 - x_4)} \cdot (x_3 - x_F) \\ &= 8 \end{aligned}$$

Equation 4.42

Hence the length BF is:-

$$BF = y_B - y_F = 1.47159090909091$$

Equation 4.43

Applying APT with Revised Data for Co-Ordinates

If we recalculate the Area of the Trapezium ADEC with the computed data for co-ordinates we are able to write:-

$$
\begin{aligned}
A_{ADEC} &= DE.EC + \frac{DE.AG}{2} \\
&= DE.\left(EC + \frac{AG}{2}\right) \\
&= (x_3 - x_4)\cdot\left[(y_2 - y_3) + \frac{(y_1 - y_2)}{2}\right] \\
&= \frac{(x_3 - x_4).(y_1 + y_2 - 2y_3)}{2} \\
&= 56
\end{aligned}
$$

Equation 4.44

Hence the new Area of the Interval, A_{coords}, is given by:-

$$A_{coords} = A_{ADEC} + A_{seg} = 63$$

Equation 4.45

The areal Percentage Specific Defect is now zero:-

$$PSD(A_{interval}, A_{coords}) = 0$$

Equation 4.46

Note again that the several interval-area estimation algorithms that we have discussed in this section are robust under interval width diversity and are thus readily iterable for the total area of a closed figure whose successive boundary co-ordinates are known.

But the key to accurate quadrature is to have as many boundary data co-ordinates as possible and to record the individual points as accurately as practicable.

If you need to fit a curve of higher degree than the second to any sequence or sub-sequence of boundary points then you have too little data. Go back into the field (literal or figurative: No pun intended and I trust none taken) and double or treble your readings.

Pixel Counting by Graphical Software

Rydal Water is a celebrated and very beautiful lake in the mountain country of the Cumbrian Lake District of Northern England. In some ways it is the British Walden Pond, lauded by poets and painters of the Romantic Era, say 1770 to 1840AD, and the resort of literary tourists to this day.

For example, the key Romantic William Wordsworth who, in "*Water-Fowl Observed Frequently Over The Lakes Of Rydal And Grasmere*", probably written in 1802 or 1803AD, gave us:-

MARK how the feathered tenants of the flood,
With grace of motion that might scarcely seem
Inferior to angelical, prolong
Their curious pastime! Shaping in mid air
(And sometimes with ambitious wing that soars
High as the level of the mountain-tops)
A circuit ampler than the lake beneath—
Their own domain; but ever, while intent
On tracing and retracing that large round,
Their jubilant activity evolves
Hundreds of curves and circlets, to and fro,
Upward and downward, progress intricate ...

I am indebted to The Brathay Exploration Group (BEG) for their fine bathymetric survey of this lake, and others in the district, and I have the honor to reproduce the resulting diagram from page 152 of their book ***"Tarns of the Central Lake District"*** (ISBN 0-90601517-0) and to present that diagram throughout this demonstration. Colorings and annotations are mine, as are all errors of interpretation.

I re-produce the Rydal Water bathymetric diagram as given in Figure 4.5, and colored by me to emphasise the bathymetry in Figure 4.6

It is clear that Rydal Water comprises a subtler Western Basin and a more abrupt and deeper Eastern Basin separated by a subaqueous ridge summited by two or three islands. Both basins are roughly the caps of spheres and it seems very likely that Rydal Water is a twin kettle generated by two remnant ice blocks set within a glacial valley. If so then the two ice blocks, the lower and Eastern bigger than the upper and Western, were separated by a hundred-meter gap or crevasse that formed the basis of the kettles' further evolution.

Note that the very center of the Western Basin is elevated with respect to the rest of the basin floor. The River Rothay enters tangentially to the lake and sets up a standing eddy in the Western Basin, a toroid of moving water with an internal structure of helical streamlines. The erosional and depositional features of this vortex are intensified in times of spate. The tendency of this current is to allow fine sediments to drift and to settle at the dead center. Thermal convection of the water may play a secondary rôle, but Coriolis effects would be negligible in such a small, shallow body of water.

Figure 4.7 illustrates the tangentiality of water access and egress at Rydal Water.

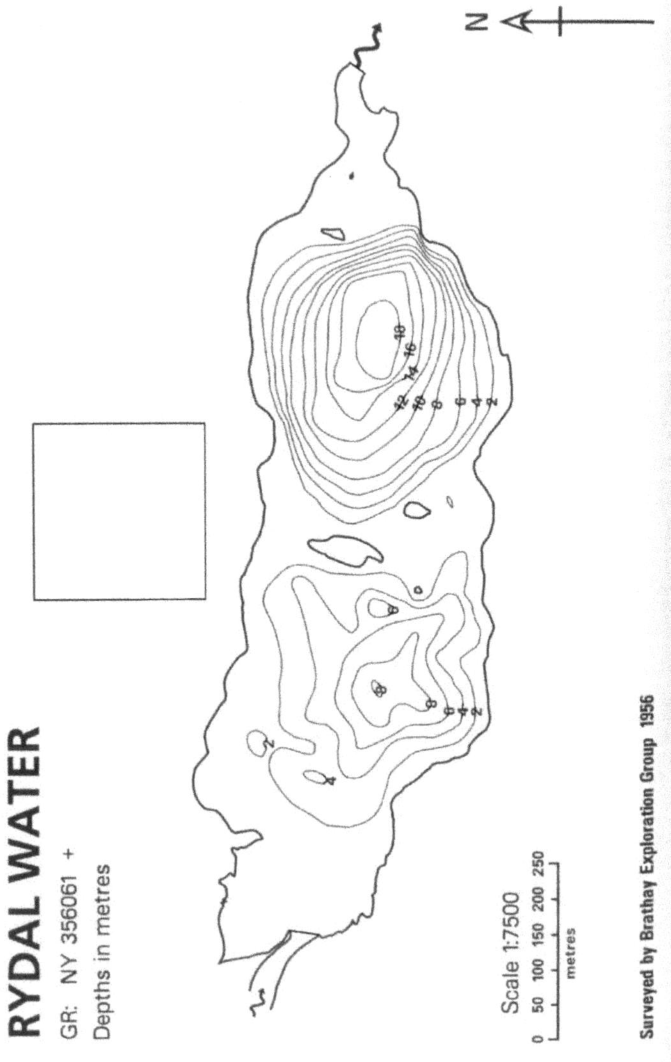

Figure 4.5
The BEG Survey of Rydal Water
with my Calibration Square
According to the Given Scale Bar

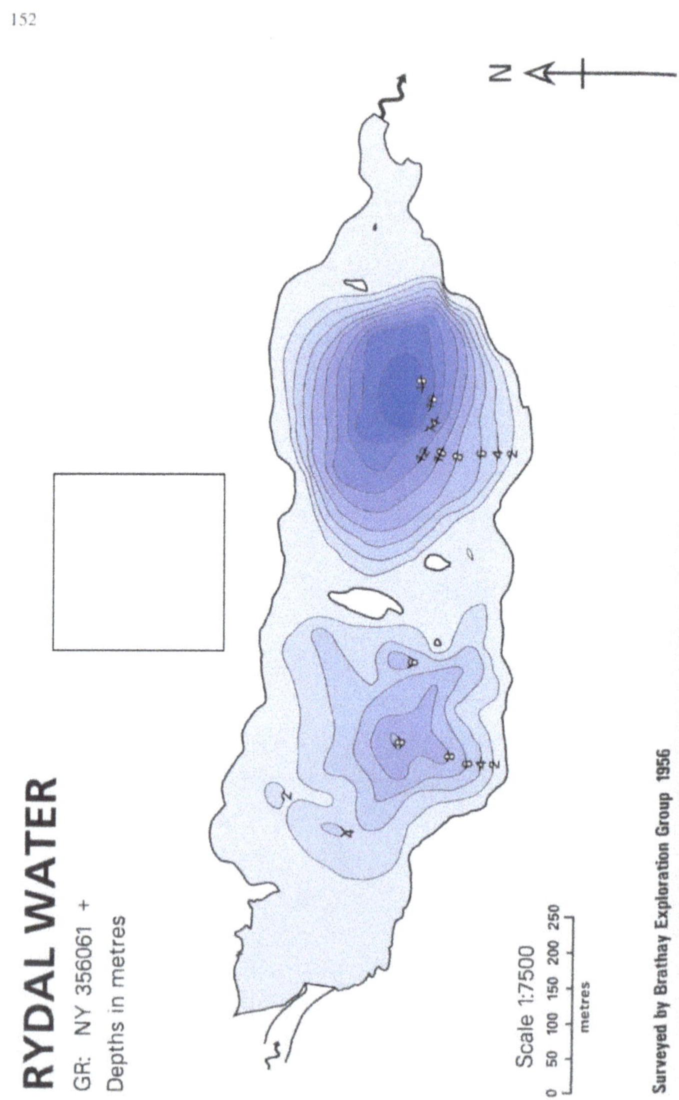

Figure 4.6
The Depths of Rydal Water

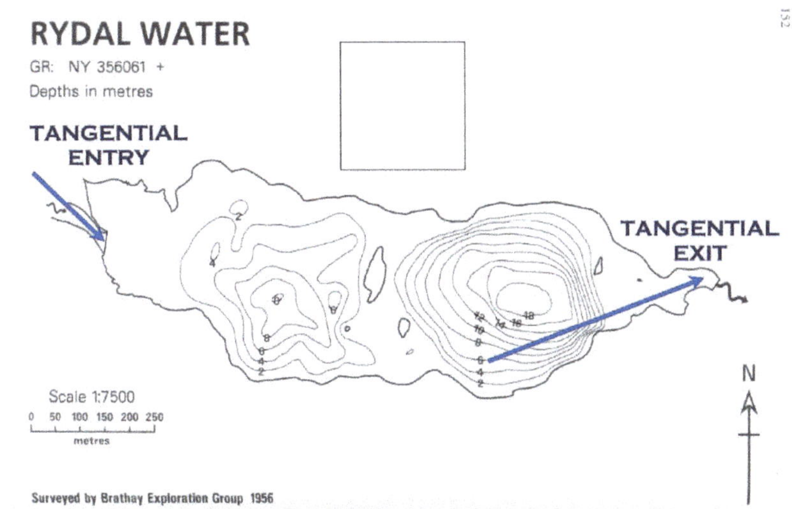

Figure 4.7
The Tangentiality of Water Flow Into and Out of Rydal Water

Using modern graphical software in conjunction with a high-quality survey such as that of Rydal Water, it is possible to estimate areas with surprising accuracy by pixel counting.

The method requires firm continuous enclosure boundaries drafted in a color contrasting with the background so that the machine system can distinguish one measurable patch from its neighbor. The BEG supplied these, interrupted only by typographical gaps for depth annotations, which I have plugged using PhotoDraw.

To count the pixels in enclosed patches I started with the 250 meter square Calibration Square which I draw at the top center of Figure 4.5, using PhotoDraw, and based, in the absence of graticulation, upon the BEG Scale Bar.

The counting utility was that provided by the graphical software package GIMP 3.0.2-1, which is superb and is downloadable free of charge.

Figure 4.8 is a screenshot of GIMP during user-piloted quadrature of Rydal Water, and is concisely annotated with the required mouse steps.

These pixel counts are summarised by Table 4.4 and the pixels per square meter (px/m^2) are computed from the pixels in the Calibration Square which is 146688 and knowledge that the square meters represented by that square are $250 \times 250 = 62500$. Accordingly there are $146688/62500 = 2.347008$ px/m^2.

Table 4.5 summarises the pixel counts on each bathymetric plane, where relevant separated for each of the Western and Eastern basins.

From Table 4.5 we can move forward to Table 4.6 "Histogram" which computes and processes isobathic plane areas in square meters.

Thirdly, Table 4.7 summarises the input data and computed outputs for the Rydal Water quadrature and volume calculation exercises.

With volume calculations we are trespassing on the purview of Cubature, the mathematical art of the computation of volumes!

Formal Cubature is beyond the scope of this book.

You can see from Table 4.7 that the $PSD(A_{fiducial},A_{infered})$ is -0.053831594, a discrepancy of about one part in eighteen hundred. This extreme accuracy (for an environmental science finding) is due to the superb surveying of Victorian Ordnance Survey officers on the one hand and The

Brathay Exploration Group volunteers on the other; and the very excellent quality of GIMP software.

The $PSD(V_{fiducial}, V_{infered})$ is +11.5282789

As you have seen from Table 4.6, I moved forward to treat the isobathic plane areas as discrete column quantities in a very basic Grouped Frequency histogramic analysis from which I assessed mean depth as the grouped mean. This is not necessarily the most realistic way of treating average lake depths, but was tailored to the situation in which only charted isobaths were available to me rather than original point soundings. The officially accepted mean depth of Rydal Water is about 5.3 meters.

Figure 4.9 is the resulting histogram with an exponential decay regression relating Depth and Area. These too are very tenuous as both the Distribution and the Regression are interpretations of interpretations.

The equation for a Grouped Mean, $x_{grouped}$, is:-

$$x_{grouped} = \frac{\sum fx}{\sum f} = \frac{\sum_{j=1}^{k}(A_j . x_j)}{\sum_{j=1}^{k} A_j}$$

Equation 4.48

where:-

$$x_j = d_j + \frac{d_{j+1} - d_j}{2}$$

Equation 4.49

d_j being the Depth of the jth Isobath of k Isobaths, and A_j its Area.

The Grouped Standard Deviation, $SD_{grouped}$, is yielded by:-

$$SD_{grouped} = \sqrt{\frac{\sum fx^2}{\sum f} - x_{grouped}{}^2}$$

$$= \sqrt{\frac{\sum_{j=1}^{k}\left(A_j . x_j{}^2\right)}{\sum_{j=1}^{k} A_j} - x_{grouped}{}^2}$$

Equation 4.50

Note that the sum of Isobath Areas, ΣA_j, is equivalent to the Lake Surface Area.

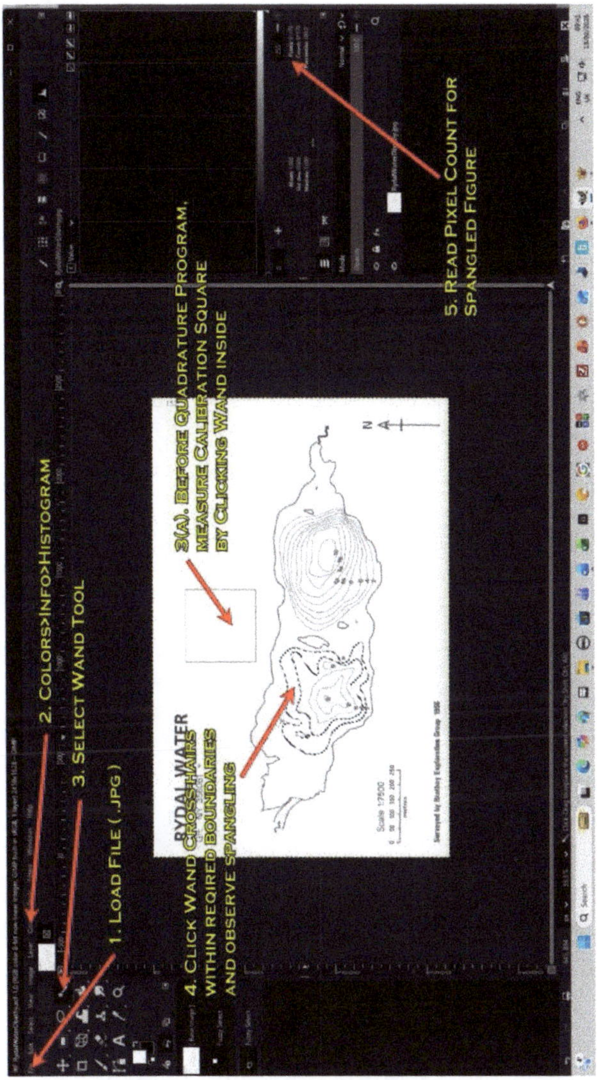

Figure 4.8
Using GIMP to Measure Relative Areas by Pixel Counting

MEASURED PIXELS					
Serial	Higher Isobath	Lower Isobath	Whole Basin	West SubBasin	East SubBasin
1	0	2	337128		
2	0	2			
3	2	4		91179	30187
4	2	4		1614	
5	4	6		48306	28418
6	4	6		732	
7	6	8		28063	27563
8	6	8		1480	
9	8	10		16458	27086
10	10	12			22929
11	12	14			25380
12	14	16			15931
13	16	18			20890
14	18	20			12407
Totals			337128	187832	210791

Table 4.5
Pixel Counts for Rydal Water

HISTOGRAM									
Serial	Higher Isobath	Lower Isobath	Abscissal Captions	Pixels	Area (m^2)	MidPoint x	fx	fx^2	
1	0	2	0-2	337128	143641.6	1	143641.6	143641.6	
2	2	4	2-4	122980	52398.63	3	157195.9	471587.7	
3	4	6	4-6	77456	33002.02	5	165010.1	825050.4	
4	6	8	6-8	57106	24331.4	7	170319.8	1192239	
5	8	10	8-10	43544	18552.98	9	166976.8	1502792	
6	10	12	10-12	22929	9769.46	11	107464.1	1182105	
7	12	14	12-14	25380	10813.77	13	140579	1827527	
8	14	16	14-16	15931	6787.791	15	101816.9	1527253	
9	16	18	16-18	20890	8900.694	17	151311.8	2572301	
10	18	20	18-20	12407	5286.305	19	100439.8	1908356	
				735751	313484.7		1404756	13152851	Sum
				73575.1	31348.47		140475.6	1315285	Mean
					4.481099				Grouped Mean
				93710.74	39927.75				SDpop
					4.677251				Grouped SD

Table 4.6
Calculation of Square Meter Areas
With
Grouped Frequency Analysis

Object Name:	Rydal Water	
Object Species:	Lake	
Location of Center:	54°26'50"N, 2°59'43"	
Surveyor:	Brathay Exploration Group	
Calibration Square Side Length:	250	meters
Calibration Square Area:	62500	meters^2
Pixels in Calibration Square:	146688	
Pixels per Square Meter:	2.347008	px/m^2
Total Counted Pixels:	735751	
Fiducial Area (Meters^2)	313316	
Infered Area (Meters^2)	313485	
PSD($A_{fiducial}$,$A_{Infered}$)	-0.053831594	
Grouped Mean Depth (meters)	4.481098904	
Fiducial Volume (m^2) (Lakes Portal DEFRA)	1587802	
Infered Volume (meters^3)	1404755.757	
PSD($V_{fiducial}$,$V_{Infered}$)	11.5282789	
Assumed Water Density (tonnes/m^3)	0.9995	
Infered Mass of Contained Water (tonnes)	1404053.379	
Software:	GIMP 3.0.2-1	

Table 4.7
Rydal Water
Statistical Summary

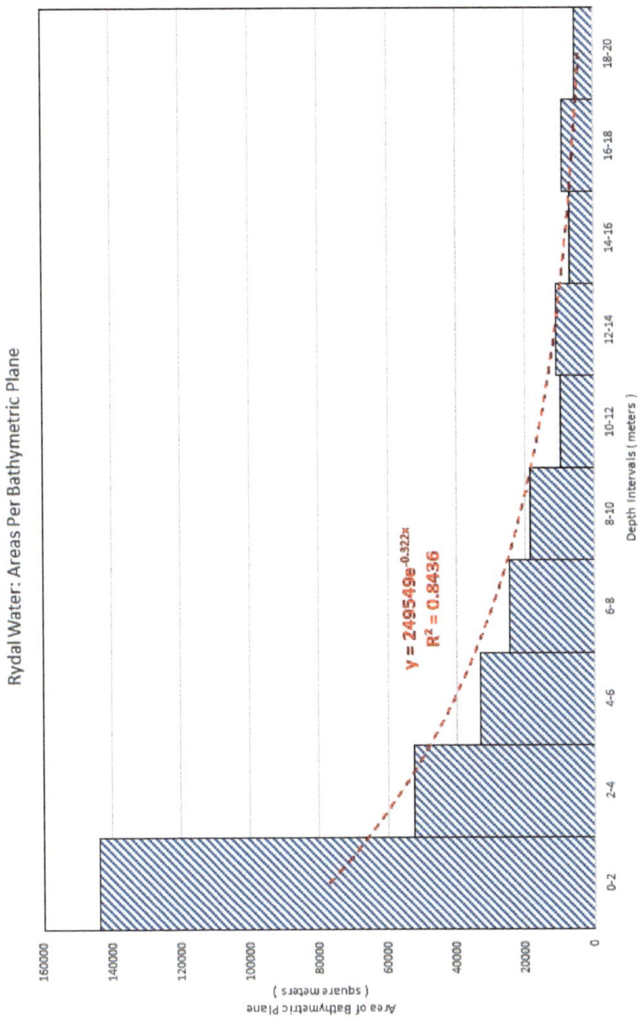

Figure 4.9
Histogram for Rydal Water
Depth Profile
with
Exponential Regression

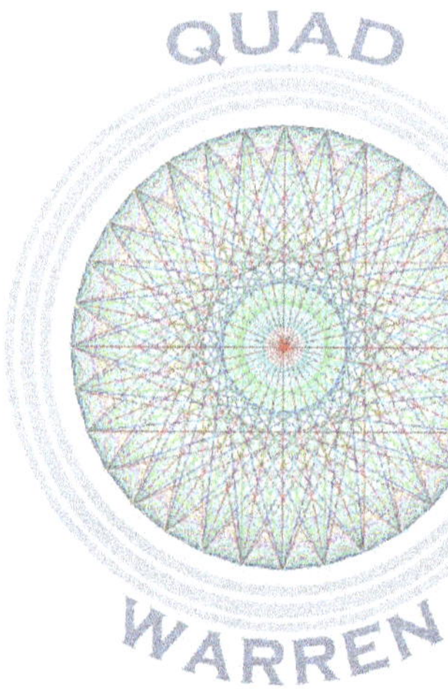
QUAD
WARREN

BIBLIOGRAPHY

Abramowitz and Stegun
"Handbook of Mathematical Functions with Formulas, Graphs, and Mathematical Tables"
Edited by Milton Abramowitz and Irene Stegun of the United States National Bureau of Standards (NBS), now the National Institute of Standards and Technology (NIST).
Published by the United States Department of Commerce
1964
ISBN 0-486-61272-4
pp 1046

NIST
"NIST Handbook of Mathematical Functions Hardback and CD-ROM"
Frank W. J. Olver (Editor), Daniel W. Lozier (Editor), Ronald F. Boisvert (Editor), Charles W. Clark (Editor)
Publisher : Cambridge University Press
Publication date : 17 May 2010
Edition : 1st
Print length : 968 pages
ISBN-13 : 978-0521192255

Bronshtein Moscow 1979
"Handbook of Mathematics"
Bronshtein-Semendyayev
Verlag Harri Deutsch of Thun 1985
Van Nostrand Reinhold Company of New York
3rd Edition
Europe: ISBN 3-87144-644-0
US: ISBN 0-442-21171-6
pp 973 Quadrature p 701, 764
See also:-

"Handbook of Mathematics"
IN Bronshtein, KA Semendyayev, Gerhard Musiol, Heiner Mühlig
Springer of Berlin and London
6th Edition 18 May 2015
ISBN 978-3662462201
1251 pp

Weisstein Conspectus
"CRC Concise Encyclopedia Mathematics"
Eric W Weisstein
2nd Edition
Chapman Hall of London and
CRC Press of Boca Raton
12 December 2002
ISBN 978-1584883470
pp 3242
Quadrature p 2437 and 2438 which re-directs to specific methods
This title is succeeded by further editions and multimedia versions

Shaum's Numerical Analysis
"Schaum's Outline of Numerical Analysis Paperback – 1 Jan. 1989"
by Francis Scheid (Author)
Publisher : McGraw-Hill
Publication date : 1 Jan. 1989
Edition : 2nd
Language : English
Print length : 480 pages
ISBN-13 : 978-0070552210

Scraton on Basic Numerical Analysis
"Basic Numerical Methods"
"An Introduction to Numerical Mathematics on a Microcomputer"
RE Scraton
Edward Arnold of London 1 September 1984
ISBN 0-7131-3521-2
pp 92

The Brathay Exploration Group on Cumbrian Tarns' Ecology and Bathymetry
"Tarns of the Central Lake District"
Elizabeth Haworth, George de Boer, Ian Evans,
Henry Osmaston, Winifred Pennington, Alan Smith
Phillip Story and Brian Ware
The Brathay Exploration Group Trust, Ambleside
2003
Charts by Philip Storey, Townend, Farleton
Carnforth, Lancashire, LA6 1PB, UK
ISBN 0-90601517-0
pp 205 pbk

The Athens Discussion of Archimedes Parabola Theorem (APT)
"3+2 THEOREMS ON PARABOLAS"
"On the mathematical prelude to Archimedes' monumental treatise Quadrature of the Parabola"
Dimitris Sardelis
Athens, April 25, 2018
Available online at Researchgate:-
https://www.researchgate.net/profile/Dimitris-Sardelis-2/publication/324746852_32_THEOREMS_ON_PARABOLAS_On_the_mathematical_prelude_to_Archimedes%27_monum

ental_treatise_Quadrature_of_the_Parabola/links/5ae061f20f7e9b285946f7cb/3-2-THEOREMS-ON-PARABOLAS-On-the-mathematical-prelude-to-Archimedes-monumental-treatise-Quadrature-of-the-Parabola.pdf
pp 11

Elementary Calculus
"Calculus"
"A Complete Introduction"
Paul Abbott and Hugh Neill
Teach Yourself Books
Hachette UK (Hodder Education division) 1992
ISBN 978-1-4441-9111-0
pp 373 pbk
Indefinate Integration of a Polynomial p 176
Differentiation p 59
Based upon the English Universities Press of London
"Teach Yourself Calculus" by Paul Abbott BA, 1940

INDEX

Appendix Locators are Not Reliable

APPENDIX ONE
SOURCE CODE AND DATA

Ellipse N=1024 (verification) QUADellipseNis1024.txt:-SOURCE CODE

```
'
' **** MASTER SEGMENT
'
'    Program QUADellipseN=1024
' A Program
' To Compute the Quadrature of an Ellipse of Area PI.a.b
' By Several Different Means which are Compared with the
' Analytic Value
' ( With check matrix of values referent to EXCEL worksheet
'   ELLIPSEquadNisABinaryExponentFhasEquations.xlsx )
' Based upon Generated Data.
'    Written by:-
'
'         James R Warren BSc MSc PhD PGCE
'         "Southgate"
'         31 Victoria Avenue
'         Bloxwich
'         Walsall
'         WS3 3HS
'
'         19 February 2026
'
'      This Program is Written in QBasic for
'      elaboration as a compiled
'      QB64 QUADellipse.exe object
'
'    Compiled QB64 on a 64-bit Dell XPS 8940
'    x64-based Personal Computer
'    Total Execution Time = Not Known
'
    Rem $DYNAMIC
```

```
' Initialise the Random Number Generator
    Randomize (Timer)
' Variable Type Defaults
    DefDbl A-H, O-R, T-Z
    DefStr S
    DefInt I-K, M-N
    DefLng L
' Declare Dynamic Array Dimensions
'    ( None )
' Assign the Dynamic Arrays' Maxima and
' Dimension the ' Dynamic Arrays
'    ( None )
' Pseudo-Static Array Declarations
'    COMMON SHARED X),Y(),TH(),R()
'    COMMON SHARED XIN(),YIN()
'    COMMON SHARED ARY(),G()
' Assign the Pseudo-Static Array Dimensions
' Maxima and Dimension the Pseudo-Static Arrays
    N = 1024
    ReDim X(N), Y(N), TH(N), R(N)
    ReDim XIN(N), YIN(N)
    ReDim ARY(N), G(N)
' Object Definitions
'    ( None )
' Public Object Instantiations
'    ( None )
' Static Array Declarations
'    ( None )
' Dynamic Array Declarations
'    ( None )
' Constant Definitions
    Const PI = 3.14159265358979
    Const PI2 = 6.28318530717959
    Const PHI = 1.618033988749890
```

```
' Declare Logical Unit Number Holders
    Common Shared ISW, IW, IX
' Declare String Constant Data
'  ( for Common Shared emulation )
    Common Shared SC, SM, SCR
' Declare Formats ( for Common Shared emulation )
    Common Shared SI4, SF10P8, SF11P6, SF15P12
    Common Shared SF13P8,    SF20P15
' Define Random Number Range Parameters
'     ( none )
' String Constant Definitions
    SC = ":": SM = ",": SCR = Chr$(13) + Chr$(10)
' Format Definitions
    SI4 = "####": SF10P8 = "#.########"
    SF11P6 = "####.######"
    SF15P12 = "##.############
    SF13P8 = "####.########"
    SF20P15 = "####.###############"
' Input File Data for Ellipse Test Job
    SON = "1024-interval Ellipse"
    Sname = "ellipse"
    SExt = ".TXT"
    Sfsin = "XYIN" + Sname + SExt
 ' Sout = "SUMMARY" + Sname + SExt
    Spath = "C:\QUAD\TXT\"
    Sconcat = Spath + Sfsin
    Print "Proceeding with "; SON
' Perform Actions
' Input Data ( for Natural Irregular Boundary jobs )
'  SUB DATAIN(SON,SINF,Sname,SMAP,SDATE,N,X(),Y())
' Local Constants Definition
    NPP = N + 1
    RANGE = PI2
    h=RANGE/N
```

```
    aaxis=(1+Sqr(5))/2
    baxis=1
    ecc=sqr(1-baxis^2/aaxis^2)
    Call CEISK(ecc,CEI)
    Afido=PI*aaxis*baxis
    Cfido=4*aaxis*CEI
' Elaboration
    Call GENERATORellipse(N, aaxis, baxis, ecc, NPP,
RANGE, dth, fsindt, TH(), TH0, THN, R(), Asec, X(),
Y(),Afido)
    Call DMEAN(N,NPP,0,N,R(),Rsum,Rmu)
    Acir=PI*Rmu^2
    Call MEISTER(N, X(), Y(), AMe)
    Call SIDES(N, X(), Y(), Capx)
    Call SIMPSONpolarType1(N, RANGE, h, R(), TH(),
Asim)
' Output and Terminate
    Call SUMMARYellipse(N, NPP, RANGE, dth, fsindt, PHI,
aaxis, baxis, ecc, CEI, TH0, THN, h, AMe, Afido, Asec, Asim,
Cfido, Capx, Rmu, Acir)
    End

'
' **** SUBSEGMENT LIBRARY
'

    Function ALT (I)
' A Function to Alternate 4 and 2 for I=1 to N-1
' for use in the Solution of the Extended Simpson's Rule
'
'  Argument:-
'     I  The Serial Counter
'        Functional Array Subscript
'

    ALT = 3 + (-1) ^ (I + 1)
```

```
    End Function

    Function PSD (x, y)
' A Function to Compute the Percentage Specific Defect
'
'  Arguments:-
'     x  The Fiducial Value
'     y  The Comparator Value
'
    PSD = 100 * ((x - y) / x)
    End Function

    Sub CEISK(ecc,CEI)
' A Subroutine to
' Establish a Convergent AGM using the deQueiroz Modus
' ( Landen's Transformation ) and from this the
' Complete Elliptic Integral of the Second Kind, CEI.
' see
https://www.robertobigoni.it/English/Matematica/Integrali/IntegraliEllittici/Elliptic.html )
'   Arguments:-
'       ecc     The Eccentricity of the Ellipse
'       CEI     The Complete Elliptic Integral of the
'               Second Kind
'
        Dk=ecc
        Dkk=Dk*Dk
        a=1
        b=sqr(1-Dkk)
        E=1-Dkk/2
        i=1
        DO WHILE ABS(a-b)>0.000000000000001
                a1=(a+b)/2
                b1=sqr(a*b)
```

```
            amb=a-b
            E=E-i*amb*amb/4
            i=i*2
            a=a1
            b=b1
        LOOP
        DIK=PI/(2*a)
        CEI=E*DIK
        End Sub

Sub CENTERISE (N, XIN(), YIN(), XD, YD)
' A Subroutine to Shift Input (X,Y) Cartesian Co-ordinates
' so that the Figure delineated is Centered at the
' Desired Center (XD, YD )
'
'  Arguments:
'     N         The Number of Sectors of Integration
'     XIN       The Array of Input Cartesian X-values
'     YIN       The Array of Input Cartesian Y-values
'     XD        The X-coordinate of the Desired Figure Center
'     YD        The Y-coordinate of the Desired  Figure Center
' Centerise X and Y-displacements
      For I = 0 To N
         XIN(I) = XIN(I) - XD
         YIN(I) = YIN(I) - YD
      Next I
' Terminate
      End Sub

      Sub DATAIN (ISW, SON, Spath, Sfile, Sname, SMAP,
SDATE, N, X(), Y())
' A Subroutine to Load the Data for Quadrature
'  Arguments:-
'     ISW    The Logical Transput Unit Number
```

```
'      SON      The Object Descriptor
'      Spath    The Directories' Pathname
'      Sfile    The Input Filename
'      Sname    The Data Name
'      SMAP     The Map Descriptor
'      SDATE    The Date of Data Compilation
'      N        The Number of Intervals
'      X()      The Array of x-Coordinates
'      Y()      The Array of y-Coordinates
'
    Call FILEOPEN(ISW, Spath, Sname, Sfile, SExt)
    Input #ISW, SON
    Input #ISW, Sfile
    Input #ISW, Sname
    Input #ISW, SMAP
    Input #ISW, SDATE
    Input #ISW, N
    For I = 1 To N
       Input #ISW, X(I), Y(I)
    Next I
' Terminate
    Close ISW
    End Sub

    Sub DMEAN(N,NPP,ILB,IUB,ARY(),T,DMU)
' A Subroutine to Sum the Array Elements of ARY()
' Between Subscripts ILB and IUB inclusive
'
'  Arguments:
'     N         The Number of Intervals
'     NPP       The Number of Interval Boundaries
'     ILB       The Lower Bound
'     IUB       The Upper Bound
'     ARY()     Local Holder Array
```

```
'      T          Local Total
'      DMU        Local Mean
'
     Call DSUM(ILB, IUB, ARY(), T)
     DMU=T/NPP
     End Sub

     Sub DSUM (ILB, IUB, ARY(), T)
' A Function to Sum the Array Elements of ARY()
' Between Subscripts ILB and IUB inclusive
'
'   Arguments:
'      ILB        The Lower Bound
'      IUB        The Upper Bound
'      ARY()      Local Holder Array
'      T          Local Total
'
     Tot = 0.0
     For I = ILB To IUB
        Tot = Tot + ARY(I)
     Next I
     T=Tot
' Terminate
     End Sub

     Sub FILEOPEN (ISW, Spath, Sname, SFile, SExt)
' A Subroutine to Open a Sequential File for Input
'   Arguments:-
'      ISW        The Logical Unit Number
'      Spath      The Directories' Path
'      Sname      The Data Name
'      SFile      The Filename
'      SExt       The Filename Extension
'
```

```
If ISW = 1 Then
   SFile = "XYIN" + Sname
   Sconcat = Spath + SFile + SExt
   Open Sconcat For Input As ISW
Else
   ISW=2
   SFile = "SUMMARY" + Sname
   Sconcat = Spath + SFile + SExt
   Open Sconcat For Output As ISW
End If
End Sub

Sub GENERATORellipse (N, aaxis, baxis, ecc, NPP, RANGE, dth, fsindt, TH(), TH0, THN, R(), Asec, X(), Y(),Afido)
' A Subroutine to Generate the N Cartesian Co-ordinates
' X(),Y() for an Ellipse of Axes aaxis and baxis
'  Arguments:-
'     N          The Number of Radial Intervals
'     aaxis      The Ellipse Major Axis Length
'     baxis      The Ellipse Minor Axis Length
'     ecc        The Eccentricity of the Ellipse
'     NPP        The Number of Perimeter Points
'     RANGE      The Total Swept Angle ( 2*PI radians )
'     dth        Incremental Change in the Swept Angle
'     fsindt     The Sine of dth
'     TH()       The Swept Angle at Point i
'     TH0        The Initial Value of Sweep Angle
'     THN        The Final   Value of Sweep Angle
'     R()        The Swept Radius at Point i
'     Asec       The Area of the Summated Sectors
'     X()        The Array of Cartesian x-Coordinates
'     Y()        The Array of Cartesian y-Coordinates
'     Afido      The Fiducial Value of Figure Area
```

```
'
        aaxis = (1 + Sqr(5)) / 2
        baxis = 1
        ecc = Sqr(1 - (baxis / aaxis) ^ 2)
        NPP = N + 1
        top = aaxis * baxis
        asq = aaxis ^ 2
        bsq = baxis ^ 2
        dth = RANGE / N
        fsindt = Sin(dth)
        For I = 0 To N
           TH(I) = I * dth
           R(I) = top / Sqr(asq * Sin(TH(I)) ^ 2 + bsq * Cos(TH(I))
^ 2)
           X(I) = R(I) * Cos(TH(I))
           Y(I) = R(I) * Sin(TH(I))
        Next I
        Asec = 0.0
        For I = 0 To N - 1
           Asec = Asec + R(I) * R(I + 1) * fsindt / 2
        Next I
        TH0 = TH(0): THN = TH(N)
        Afido=PI*aaxis*baxis
' Terminate
End Sub

        Sub GRAVICENTER (N, XIN(), YIN(), XG, YG)
' A Subroutine to Compute the Gravitational Center of a Figure
' from its Cartesian Boundary Co-ordinates XIN() and YIN()
'
'   Arguments:
'      N    The Number of Sectors of Integration
'      XIN The Array of Input Cartesian X-values
'      YIN The Array of Input Cartesian Y-values
```

```
'     XG  The X-coordinate of the Gravimetric Figure Center
'     YG  The Y-coordinate of the Gravimetric Figure Center
' Zeroise Counters
      TX = 0
      TY = 0
' Total X and Y-displacements
      For I = 0 To N
         TX = TX + XIN(I)
         TY = TY + YIN(I)
      Next I
' Establish Cartesian Gravimetric Center
      XG = TX / N
      YG = TY / N
' Terminate
      End Sub

      Sub MEISTER (N, X(), Y(), AMe)
' A Subroutine to Compute the Meister's Rule
' Area from the Cartesian Co-ordinates X() and Y()
'
'   Arguments:-
'     N          The Number of ( Equal ) Intervals
'                of Array X()
'     X()        The Array of Values of x
'     Y()        The Array of Values of y
'     AMe        The Meister's Rule Area
'
      AMe = 0.0
      For I = 1 To N - 1
         AMe = AMe + Y(I) * (X(I - 1) - X(I + 1))
      Next I
      AMe=0.5*AMe
' Terminate
      End Sub
```

```
    Sub SIDES (N, X(), Y(), Capx)
' A Subroutine to Estimate the Perimeter of an Ellipse by
' Summing its Side Segments
'  Arguments:-
'
'     N          The Number of Intervals of Array X()
'     X()        The Array of Values of x
'     Y()        The Array of Values of y
'     Capx       The Sum of Side Segments
'
    Capx = 0.0
    For I = 0 To N - 1
      DX = X(I + 1) - X(I)
    DY = Y(I + 1) - Y(I)
    Capx = Capx + Sqr(DX ^ 2 + DY ^ 2)
    Next
' Terminate
    End Sub

    Sub SIMPSONcartesian (N, RANGE, h,X(), Y(), ILB, IUB,
Asim)
' A Subroutine to Perform Simpson's Rule
' Numerical Integration upon the Array Y()
' of the Subject Abscissal Function
'
'  Arguments:-
'     N          The Number of ( Equal ) Intervals
'                of Array X()
'     RANGE  The Total Swept Angle ( 2*PI radians )
'     h          The Interval Width
'     X()        The Array of Values of x
'     Y()        The Array of Functions of x to be
'                Integrated
```

```
'     ILB        The Range Lower Bound
'     IUB        The Range Upper Bound
'     Asim       The Simpson Integral
'

    h = (X(N) - X(0)) / N
    TT = 0.0
    For I = 1 To N - 1
       TT = ALT(I) * Y()
    Next I
    Asim = (h / 3) * (Y(0) + TT + Y(N))
' Terminate
    End Sub

    Sub SIMPSONpolarType1 (N, RANGE, h, R(), TH(),
Asim)
' A Subroutine to Perform Simpson's Rule
' Numerical Integration upon the Array Y()
' of the Subject Abscissal Function.
' The Method is Adjusted for Polar Co-Ordinates.
'
'   Arguments:-
'     N          The Number of ( Equal ) Intervals
'                of Array TH()
'     RANGE      The Total Swept Angle ( 2*PI radians )
'     h          The Interval Width
'     R()        The Array of Values of Swept Radius
'     TH()       The Array of Values of Incremental Angles
'     Asim       The Simpson Integral
'

    h=RANGE/N
    G0 = R(0) ^ 2
    GN = R(N) ^ 2
    TT = 0.0
    For I = 1 To N - 1
```

```
        TT = TT + ALT(I) * R(I) ^ 2
      Next I
      TT = G0 + TT + GN
      Asim = (1 / 6) * ((TH(N) - TH(0)) / N) * TT
' Terminate
      End Sub

      Sub SIMPSONpolarType2 (N, RANGE,h,R(), TH(), ILB,
IUB, Asim)
' A Subroutine to Perform Simpson's Rule
' Numerical Integration upon the Array Y()
' of the Subject Abscissal Function.
' The Method is Adjusted for Polar Co-Ordinates.
'
'   Arguments:-
'     N          The Number of ( Equal ) Intervals
'                of Array TH()
'     RANGE      The Total Swept Angle ( 2*PI radians )
'     h          The Interval Width
'     R()        The Array of Values of Swept Radius
'     TH()       The Array of Values of Incremental Angles
'     ILB        The Range Lower Bound
'     IUB        The Range Upper Bound
'     Asim       The Simpson Integral
'

      h = RANGE/N
      Dim G(1024)
      G(0) = R(0) ^ 2
      G(N) = R(N) ^ 2
      For I = 1 To N - 1
         G(I) = 0.0
         G(I) = R(I) ^ 2
      Next I
```

```
    Call SIMPSONcartesian(N, RANGE, h, TH(), G(), ILB,
IUB, Asim)
' Terminate
    End Sub

    Sub SUMMARYellipse (N, NPP, RANGE, dth, fsindt, PHI,
aaxis, baxis, ecc, CEI, TH0, THN, h, AMe, Afido, Asec, Asim,
Cfido, Capx, Rmu, Acir)
' A Subroutine to Output the Summary Data and Results to
' File SUMMARYellipse.CSV on Logical Unit ISW=2
'
'  Arguments:-
'     N         The Number of Radial Intervals
'     NPP       The Number of Perimeter Points
'     RANGE     The Angular Range of the Integrative Sweep
'     dth       Incremental Change in the Swept Angle
'     fsindt    The Sine of dth
'     PHI       The Major Ratio of Phidias
'     aaxis     The Major Axis of the Ellipse
'     baxis     The Minor Axis of the Ellipse
'     ecc       The Ellipse Eccentricity
'     CEI       The Elliptic Integral of the Second Kind
'     TH0       The Initial Value of Sweep Angle
'     THN       The Final   Value of Sweep Angle
'     h         The Simpson Interval
'     AMe       The Figure Area by Meister's Rule
'     Afido     The Fiducial Figure Area PI*2*aaxis*2*baxis
'     Asec      The Figure Area by Sector Summation
'     Asim      The Figure Area by the Polar Form of
'               Simpson's Rule
'     Cfido     The Fiducial Ellipse Perimeter 4*a*ECEI
'     Capx      The Approximate Ellipse Perimeter
'     Rmu       The Radius of the Reference Circle
'     Acir      The Area    of the Reference Circle
```

```
'
' Open the File
    ISW = 2
    SOUTpath = "D:\QUAD\BAS\"
    Sname = "SUMMARYellipse"
    SExt = ".CSV"
    OPEN "D:\QUAD\BAS\SUMMARYellipseN=1024.CSV"
FOR OUTPUT AS 2
'   Call FILEOPEN(ISW, SOUTpath, Sname, SFile, SExt)
' Output the Data
    Write #ISW, "The Number of Radial Intervals", N
    Write #ISW, "The Number of Perimeter Points", NPP
    Write #ISW, "The Angular Range of the Integrative
Sweep", RANGE
    Write #ISW, "Incremental Change in the Swept Angle", dth
    Write #ISW, "The Sine of dth", fsindt
    Write #ISW, "The Major Ratio of Phidias", PHI
    Write #ISW, "The Major Semi-Axis of the Ellipse", aaxis
    Write #ISW, "The Minor Semi-Axis of the Ellipse", baxis
    Write #ISW, "The Ellipse Eccentricity", ecc
    Write #ISW, "The Elliptic Integral of the Second Kind",
CEI
    Write #ISW, "The Initial Value of Sweep Angle", TH0
    Write #ISW, "The Final  Value of Sweep Angle", THN
    Write #ISW, "The Simpson Interval", h
    Write #ISW,
    Write #ISW, "The Mean Radius", Rmu
    Write #ISW, "The Reference Circle Area", Acir
    Write #ISW, "The Figure Area by Meister's Rule", AMe
    Write #ISW,
    Write #ISW, "The Fiducial Figure Area PI*aaxis*baxis",
Afido
    Write #ISW, "The Figure Area by Sector Summation", Asec
```

```
    Write #ISW, "The Figure Area by the Polar Form of
Simpson's Rule", Asim
    Write #ISW,
    Write #ISW, "PSD(Afido,Asec)", PSD(Afido, Asec)
    Write #ISW, "PSD(Afido,Asim)", PSD(Afido, Asim)
    Write #ISW, "PSD(Afido,AMe)", PSD(Afido, AMe)
    Write #ISW, "PSD(Afido,Acir)", PSD(Afido, Acir)
    Write #ISW,
    Write #ISW, "The Fiducial Ellipse Perimeter 4*a*CEI",
Cfido
    Write #ISW, "The Approximate Ellipse Perimeter", Capx
    Write #ISW, "PSD(Cfido,Capx)", PSD(Cfido, Capx)
    Write #ISW,
    Radfido=Sqr(Afido/PI)
    AreaRefCir=PI*Radfido^2
    Write #ISW, "The Radius of the Reference Circle", Radfido
    Write #ISW, "The Area    of the Reference Circle",
AreaRefCir
    Write #ISW,
' Terminate
    Close ISW
    End Sub
```

Ellipse N=1024 (verification) SUMMARYellipse=1024.csv:- OUTPUT

The Number of Radial Intervals	1024
The Number of Perimeter Points	1025
The Angular Range of the Integrative Sweep	6.283185307179590
Incremental Change in the Swept	0.006135923151543
The Sine of dth	0.006135884649154
The Major Ratio of Phidias	1.618033988749890
The Major Semi-Axis of the Ellipse	1.618033988749890
The Minor Semi-Axis of the Ellipse	1
The Ellipse Eccentricity	0.786151378000000
The Elliptic Integral of the Second Kind	1.288568295000000
The Initial Value of Sweep Angle	0
The Final Value of Sweep Angle	6.283185307179590
The Simpson Interval	0.006135923151543
The Mean Radius	1.254205855403060
The Reference Circle Area	4.941826804647460
The Figure Area by Meister's Rule	5.083159834667730
The Fiducial Figure Area PI*aaxis*baxis	5.083203692315250
The Figure Area by Sector Summation	5.083159834667750
The Figure Area by the Polar Form of Simpson's Rule	5.083203692315260
PSD(Afido,Asec)	0.000862795397539

PSD(Afido,Asim)	-0.000000000000105
PSD(Afido,AMe)	0.000862795397783
PSD(Afido,Acir)	2.781255606214020
The Fiducial Ellipse Perimeter 4*a*CEI	8.339789189347800
The Approximate Ellipse Perimeter	8.339765962128620
PSD(Cfido,Capx)	0.000278510867023
The Radius of the Reference Circle	1.272019649514070
The Area of the Reference Circle	5.083203692315250

Bean Curve N=1024 (verification) QUADbeancurveN=1024.txt:- SOURCE CODE

```
'
' **** MASTER SEGMENT
'
'    Program QUADbeancurveN=1024
' A Program
' To Compute the Quadrature of a Bean Curve
' By Several Different Means which are Compared with the
' PseudoAnalytic Value
' ( With check matrix of values referent to EXCEL worksheet
'   BEANCURVEquadNisABinaryExponentC.xlsx )
' Based upon Generated Data
'    Written by:-
'
'         James R Warren BSc MSc PhD PGCE
'         "Southgate"
'         31 Victoria Avenue
'         Bloxwich
'         Walsall
'         WS3 3HS
'
'         25 February 2026
'
'    This Program is Written in QBasic for elaboration
'    as a compiled
'    QB64 QUADbeancurve.exe object
'
'    Compiled QB64 on a 64-bit Dell XPS 8940
'    x64-based Personal Computer
'    Total Execution Time = Not Known
'
     Rem $DYNAMIC
```

```
' Initialise the Random Number Generator
     Randomize (Timer)
' VARIABLE TYPE DEFAULTS
     DefDbl A-H, O-R, T-Z
     DefStr S
     DefInt I-K, M-N
     DefLng L
' Declare Dynamic Array Dimensions
'    ( None )
' Assign the Dynamic Arrays' Maxima and Dimension
' the Dynamic Arrays
'    ( None )
' Pseudo-Static Array Declarations
'    COMMON SHARED X(),Y(),TH(),R()
'    COMMON SHARED XIN(),YIN()
'    COMMON SHARED ARY(),G()
' Assign the Pseudo-Static Array Dimensions Maxima
' and Dimension the Pseudo-Static Arrays
     N = 1024
     ReDim X(N), Y(N), TH(N), R(N)
     ReDim XIN(N), YIN(N)
     ReDim ARY(N), G(N)
' Object Definitions
'    ( None )
' Public Object Instantiations
'    ( None )
' Static Array Declarations
'    ( None )
' Dynamic Array Declarations
'    ( None )
' Constant Definitions
'    ( None )
' Declare Logical Unit Number Holders
     Common Shared ISW, IW, IX
```

```
' Declare String Constant Data
' ( for Common Shared emulation )
    Common Shared SC, SM, SCR
' Declare Formats ( for Common Shared emulation )
    Common Shared SI4, SF10P8, SF11P6,
    Common Shared SF15P12, SF13P8, SF20P15
' Define Random Number Range Parameters
'     ( none )
' String Constant Definitions
    SC = ":": SM = ",": SCR = Chr$(13) + Chr$(10)
' Format Definitions
    SI4 = "####": SF10P8 = "#.########"
    SF11P6 = "####.######"
    SF15P12 = "##.############"
    SF13P8 = "####.########"
    SF20P15 = "####.###############"
' Preliminary Definitions
    COMMON SHARED PI, PI2
    ISW=2
    PI=3.141592653589793
    PI2=PI*2
    AA=1
    Call
GENERATORbeancurve(N,AA,X(),Y(),TH(),R(),xbar,ybar,Af
ido,Asect,Asimp,Cfido,Csegm,dth,fsindth,TH0,THN,h,Rmu,A
circ)
    Call MEISTER (N, X(), Y(), Ameis)
    Call
SUMMARYbeancurve(N,AA,xbar,ybar,Afido,Ameis,Asect,As
imp,Cfido,Csegm,dth,fsindth,TH0,THN,h,Rmu,Acirc)
    End

'
' **** SUBSEGMENT LIBRARY
```

```
'
      Function PSD (x, y)
' A Function to Compute the Percentage Specific Defect
'
'  Arguments:-
'     x  The Fiducial Value
'     y  The Comparator Value
'
      PSD = 100 * ((x - y) / x)
      End Function

      Sub FILEOPEN (ISW, Spath, Sname, SFile, SExt)
' A Subroutine to Open a Sequential File for Input
'  Arguments:-
'     ISW       The Logical Unit Number
'     Spath     The Directories' Path
'     Sname     The Data Name
'     SFile     The Filename
'     SExt      The Filename Extension
'
      If ISW = 1 Then
          SFile = "XYIN" + Sname
          Sconcat = Spath + SFile + SExt
          Open Sconcat For Input As ISW
      Else
         ISW=2
         SFile = "SUMMARY" + Sname
         Sconcat = Spath + SFile + SExt
         Open Sconcat For Output As ISW
      End If
      End Sub

      Sub
GENERATORbeancurve(N,AA,X(),Y(),TH(),R(),xbar,ybar,Af
```

```
ido,Asect,Asimp,Cfido,Csegm,dth,fsindth,TH0,THN,h,Rmu,A
circ)
' A Subroutine to Compute the Characteristics of a Bean Curve
'   Arguments:-
'       N       The Number of Radial Intervals
'       AA      The Bean Curve Axis
'       X()     The Array of Boundary x-Coordinates
'       Y()     The Array of Boundary y-Coordinates
'       TH()    The Array of Values of
'                Incremental Angles
'       R()     The Array of Values of Swept Radius
'       xbar    The Geometric Centroid x-Coordinate
'       ybar    The Geometric Centroid y-Coordinate
'       Afido   The Fiducial Bean Curve Area
'       Asect   The Sum of Sectors Bean Curve Area
'       Asimp   The Simpson's Rule Bean Curve Area
'       Cfido   The Fiducial Bean Curve Perimeter
'       Csegm   The Bean Curve Perimeter by
'               Segment Summation
'
'       dth     The Increment of the Swept Angle
'       fsindth The Sine of the Increment of the
'                Swept Angle
'       TH0     The Initial Sweep Angle
'       THN     The Final   Sweep Angle
'       h       The Simpson Increment
'       Rmu     The Mean Reference Radius
'       Acirc   The Reference Circle Area
'
' Establish Boundary Points
    dth=PI/N:fsindth=sin(dth)
    xbar=AA*(3/10): ybar=AA*(3/10)
    FOR I=0 to N
      DSN=0.0:DCO=0.0
```

```
            TH(I)=I*dth
            DSN=sin(TH(I))
            DCO=cos(TH(I))
            R(I)=AA*(DSN^3+DCO^3)
            X(I)=R(I)*DCO
            Y(I)=R(I)*DSN
        NEXT I
        TH0=TH(0)
        THN=TH(N)
' Compute Area Parameters
        Afido=(5/16)*PI*AA^2
'   Approximation Phases
        Tot1=0.0
        FOR I=0 TO N-1
            Tot1=Tot1+R(I)*R(I+1)*fsindth
        NEXT I
        Asect=0.5*Tot1
        Tot2=0.0
        FOR I=0 TO N-1
            Tot2=Tot2+(3+(-1)^(I-1))*R(I)^2
        NEXT I
        Asimp=(dth/6)*(R(0)+Tot2+R(N))
' Compute Perimeter Outputs
        Cfido=3.9317379928553
        Csegm=0.0
        FOR I=1 to N
            Csegm=Csegm+sqr((X(I)-X(I-1))^2+(Y(I)-Y(I-1))^2)
        NEXT I
' Compute Auxiliary Variables
        dth=(TH(N)-TH(0))/N: fsindth=sin(dth)
        h=dth
        Rmu=sqr(Afido/PI)
        Acirc=PI*Rmu^2
        End Sub
```

```
    Sub MEISTER (N, X(), Y(), AMe)
' A Subroutine to Compute the Meister's Rule
' Area from the Cartesian Co-ordinates X() and Y()
'
'  Arguments:-
'     N          The Number of ( Equal ) Intervals
'                of Array X()
'     X()        The Array of Values of x
'     Y()        The Array of Values of y
'     AMe        The Meister's Rule Area
'
    AMe = 0.0
    For I = 1 To N - 1
       AMe = AMe + Y(I) * (X(I - 1) - X(I + 1))
    Next I
    AMe=0.5*AMe
' Terminate
    End Sub

    Sub
SUMMARYbeancurve(N,AA,xbar,ybar,Afido,Ameis,Asect,As
imp,Cfido,Csegm,dth,fsindth,TH0,THN,h,Rmu,Acirc)
' A Subroutine to Output the Summary Data and Results to
' File SUMMARYbeancurve.CSV on Logical Unit ISW=2
'
'  Arguments:-
'        N       The Number of Radial Intervals
'        AA      The Bean Curve Axis
'        xbar    The Geometric Centroid x-Coordinate
'        ybar    The Geometric Centroid y-Coordinate
'        Afido   The Fiducial Bean Curve Area
'        Ameis   The Meister's Rule Area
'        Asect   The Sum of Sectors Bean Curve Area
```

```
'       Asimp The Simpson's Rule Bean Curve Area
'       Cfido  The Fiducial Bean Curve Perimeter
'       Csegm The Bean Curve Perimeter by
'                Segment Summation
'
'       dth     The Increment of the Swept Angle
'       fsindth The Sine of the Increment of the Swept Angle
'       TH0    The Initial Sweep Angle
'       THN    The Final   Sweep Angle
'       h       The Simpson Increment
'       Rmu    The Mean Reference Radius
'       Acirc  The Reference Circle Area
'
' Open the File
    ISW = 2
    SOUTpath = "D:\QUAD\BAS\"
    Sname = "SUMMARYbeancurve"
    SExt = ".CSV"
    OPEN
"D:\QUAD\BAS\SUMMARYbeancurveN=1024.CSV"    FOR
OUTPUT AS 2
'   Call FILEOPEN(ISW, SOUTpath, Sname, SFile, SExt)
' Output the Data
    Write #ISW, "The Number of Radial Intervals",  N
    Write #ISW, "The Number of Perimeter Points",N+1
    Write #ISW, "The Bean Curve Axia",  AA
    Write #ISW, "The Bean Curve Fiducial Area", Afido
    Write #ISW, "The Bean Curve Fiducial Perimeter", Cfido
    Write #ISW, "The Geometric Centroid x-Coordinate",  xbar
    Write #ISW, "The Geometric Centroid y-Coordinate",  ybar
    Write #ISW,
    Write #ISW, "Range (PI)",PI
    Write #ISW, "Increment of Swept Angle",dth
    Write #ISW, "Sine of Swept Angle Increment", fsindth
```

```
    Write #ISW, "Initial Swept Angle", TH0
    Write #ISW, "Final   Swept Angle", THN
    Write #ISW, "Simpson Interval",h
    Write #ISW,
    Write #ISW, "The Mean Radius", Rmu
    Write #ISW, "The Reference Circle Area", Acirc
    Write #ISW, "The Figure Area by Meister's Rule", Ameis
    Write #ISW,
    Write #ISW, "The Fiducial Figure Area", Afido
    Write #ISW, "The Figure Area by Sector Summation",
Asect
    Write #ISW, "The Figure Area by the Polar Form of
Simpson's Rule", Asimp
    Write #ISW,
    Write #ISW, "PSD(Afido,Asect)",PSD(Afido,Asect)
    Write #ISW, "PSD(Afido,Asimp)",PSD(Afido,Asimp)
    Write #ISW, "PSD(Afido,Ameis)",PSD(Afido,Ameis)
    Write #ISW, "PSD(Afido,Acirc)",PSD(Afido,Acirc)
    Write #ISW,
    Write #ISW, "The Fiducial    Figure Perimeter", Cfido
    Write #ISW, "The Approximate Figure Perimeter", Csegm
    Write #ISW, "PSD(Cfido,Csegm)", PSD(Cfido, Csegm)
    Write #ISW,
    Radfido=Sqr(Afido/PI)
    AreaRefCir=PI*Radfido^2
    Write #ISW, "The Radius of the Reference Circle", Radfido
    Write #ISW, "The Area     of the Reference Circle",
AreaRefCir
    Write #ISW, "Range of the Reference Circle (2*PI)",PI2
    Write #ISW, "Reference Circle Angular Increment", PI2/N
    Write #ISW,
' Terminate
    Close ISW
    End Sub
```

Bean Curve N=1024 (verification) SUMMARYbeancurveN=1024.CSV:- OUTPUT

The Number of Radial Intervals	1024
The Number of Perimeter Points	1025
The Bean Curve Axia	1
The Bean Curve Fiducial Area	0.981747704246810
The Bean Curve Fiducial Perimeter	3.931737992855300
The Geometric Centroid x-Coordinate	0.3
The Geometric Centroid y-Coordinate	0.3
Range (PI)	3.141592653589790
Increment of Swept Angle	0.003067961575771
Sine of Swept Angle Increment	0.003067956762966
Initial Swept Angle	0
Final Swept Angle	3.141592653589790
Simpson Interval	0.003067961575771
The Mean Radius	0.559016994374947
The Reference Circle Area	0.981747704246810
The Figure Area by Meister's Rule	0.981737847663292
The Fiducial Figure Area	0.981747704246810
The Figure Area by Sector Summation	0.981737847663291
The Figure Area by the Polar Form of Simpson's Rule	0.981747704246810
PSD(Afido,Asect)	0.001003983352972
PSD(Afido,Asimp)	0.000000000000034
PSD(Afido,Ameis)	0.001003983352881
PSD(Afido,Acirc)	-0.000000000000011

The Fiducial Figure Perimeter	3.931737992855300
The Approximate Figure Perimeter	3.931692188440440
PSD(Cfido,Csegm)	0.001164991536438
The Radius of the Reference Circle	0.559016994374947
The Area of the Reference Circle	0.981747704246810
Range of the Reference Circle (2*PI)	6.283185307179580
Reference Circle Angular Increment	0.006135923151543

Ellesmere (Figure) N=80 QUADfigureN=80.txt:- SOURCE CODE

```
'
' **** MASTER SEGMENT
'
'     Program QUADfigureN=80
' A Program
' To Compute the Quadrature of Ellesmere,
' a lake of the Salopian Lake District
' By Several Different Means which are Compared with the
' Value given ( in acres ) by the Ordnance Survey
' ( With check matrix of values referent to EXCEL worksheet
'   LAKERADIALS.xlsx )
' Based upon Boundary Data Taken from Digital Map Facsimiles
' by
' James R Warren
' using PhotoDraw and PlotDigitizer software.
'     Written by:-
'
'          James R Warren BSc MSc PhD PGCE
'          "Southgate"
'          31 Victoria Avenue
'          Bloxwich
'          Walsall
'          WS3 3HS
'
'          28 February 2026
'
'     This Program is Written in QBasic for
'     elaboration as a compiled
'     QB64 QUADfigure.exe object
'
'     Compiled QB64 on a 64-bit Dell XPS 8940
```

```
'     x64-based Personal Computer
'     Total Execution Time = Not Known
'
      Rem $DYNAMIC
' Initialise the Random Number Generator
      Randomize (Timer)
' VARIABLE TYPE DEFAULTS
      DefDbl A-H, O-R, T-Z
      DefStr S
      DefInt I-K, M-N
      DefLng L
' Declare Dynamic Array Dimensions
'     ( None )
' Assign the Dynamic Arrays' Maxima and
' Dimension the Dynamic Arrays
'     ( None )
' Pseudo-Static Array Declarations
'     COMMON SHARED X),Y(),TH(),R()
'     COMMON SHARED XIN(),YIN()
'     COMMON SHARED ARY(),G()
' Assign Fiducial Environmental Constants
'   Afido=460735
'   Cfido=3000
' Assign Desired Center Co-Ordinates
'   XD=905.0477958
'   YD=899.9775
' Assign the Pseudo-Static Array Dimensions Maxima and
' Dimension the Pseudo-Static Arrays
      N = 80: NP=N+1
      ReDim X(N), Y(N), TH(N), R(N)
      ReDim XIN(N), YIN(N)
      ReDim ARY(N), G(N)
' Object Definitions
'     ( None )
```

```
' Public Object Instantiations
'     ( None )
' Static Array Declarations
'     ( None )
' Dynamic Array Declarations
'     ( None )
' Constant Definitions
      Const PI = 3.14159265358979
      Const PI2 = 6.28318530717959
      Const PHI = 1.618033988749890
' Declare Logical Unit Number Holders
      Common Shared ISW, IW, IX
' Declare String Constant Data
' ( for Common Shared emulation )
      Common Shared SC, SM, SCR
' Declare Formats ( for Common Shared emulation )
      Common Shared SI4, SF10P8, SF11P6,
      Common Shared SF15P12, SF13P8, SF20P15
' Define Random Number Range Parameters
'     ( none )
' String Constant Definitions
      SC = ":": SM = ",": SCR = Chr$(13) + Chr$(10)
' Format Definitions
      SI4 = "####": SF10P8 = "#.########"
      SF11P6 = "####.######"
      SF15P12 = "##.############"
      SF13P8 = "####.########"
      SF20P15 = "####.###############"
' Input File Data for Ellipse Test Job
     SON = "80-interval Ellesmere"
     Sname = "Ellesmere"
     SExt = ".DAT"
     Sfsin = "XYIN" + Sname + SExt
' Sout = "SUMMARY" + Sname + SExt
```

```
    Spath = "D:\QUAD\BAS\"
    Sconcat = Spath + Sfsin
    Print "Proceeding with "; SON
' Local Constants Definition
    NPP = N + 1
    RANGE = PI2
    h=RANGE/N
' Elaboration
    Call DATAIN (1, SON, Spath, Sfile, Sname, SMAP,
SDATE, Afido, Cfido, XD,YD,N, XIN(), YIN())
    Call GRAVICENTER (N, XIN(), YIN(), XG, YG)
    Call CENTERISE (N, XIN(), YIN(), X(), Y(), XD, YD)
'   Call PRINTXY(N,X(),Y())
    Call
OMNIBUS(N,X(),Y(),TH(),R(),XG,YG,RANGE,Afido,Asect,
Asimp,Cfido,Csegm,dth,fsindth,TH0,THN,h,Rmu,Acirc)
'   Call DMEAN(N,NPP,0,N,R(),Rsum,Rmu)
    Call MEISTER(N, X(), Y(), Ameis)
'   Call SIDES(N, X(), Y(), Csegm)
    Call SIMPSONpolarType1(N, RANGE, h, R(), TH(),
Asimp)
' Output and Terminate
    Call SUMMARYfigure (2,SON, SFile, Sname, SMAP,
SDATE,N, NPP, RANGE, XG, YG, XD, YD,dth, fsindth,TH0,
THN, h, Ameis, Afido, Asect, Asimp, Cfido, Csegm, Rmu,
Acirc)
' Terminate
    End


'
' **** SUBSEGMENT LIBRARY
'

    Function ALT (I)
' A Function to Alternate 4 and 2 for I=1 to N-1
```

```
' for use in the Solution of the Extended Simpson's Rule
'
'  Argument:-
'     I  The Serial Counter
'        Functional Array Subscript
'
     ALT = 3 + (-1) ^ (I + 1)
' Terminate
     End Function

     Function PSD (x, y)
' A Function to Compute the Percentage Specific Defect
'
'  Arguments:-
'     x  The Fiducial Value
'     y  The Comparator Value
'
     PSD = 100 * ((x - y) / x)
' Terminate
     End Function

     Sub CENTERISE (N, XIN(), YIN(), X(), Y(), XD, YD)
' A Subroutine to Shift Input (XIN,YIN) Cartesian Co-ordinates
' so that the Figure delineated is Centered at the
' Desired Center (XD, YD ).
' The Output Coordinates are X(), Y().
'
'  Arguments:
'     N         The Number of Sectors of Integration
'     XIN()     The Array of Input  Cartesian X-values
'     YIN()     The Array of Input  Cartesian Y-values
'     X()       The Array of Output Cartesian X-values
'     Y()       The Array of Output Cartesian Y-values
'     XD        The X-coordinate of the Desired Figure Center
```

```
'      YD          The Y-coordinate of the Desired  Figure Center
' Centerise X and Y-displacements
     For I = 0 To N
        X(I) = XIN(I) - XD
        Y(I) = YIN(I) - YD
     Next I
' Terminate
     End Sub

     Sub DATAIN (ISW, SON, Spath, Sfile, Sname, SMAP,
SDATE, Afido, Cfido, XD,YD,N, X(), Y())
' A Subroutine to Load the Data for Quadrature
'   Arguments:-
'      ISW       The Logical Transput Unit Number
'      SON       The Object Descriptor
'      Spath     The Directories' Pathname
'      Sfile     The Input Filename
'      Sname     The Data Name
'      SMAP      The Map Descriptor
'      SDATE     The Date of Data Compilation
'      Afido     The Fiducial Figure Area
'      Cfido     The Fiducial Figure Perimeter
'      XD        The Desired Center x-Coordinate
'      YD        The Desired Center y-Coordinate
'      N         The Number of Intervals
'      X()       The Array of x-Coordinates
'      Y()       The Array of y-Coordinates
'
'   Call FILEOPEN(ISW, Spath, Sname, Sfile, SExt)
      OPEN     "D:\QUAD\BAS\XYINEllesmere.CSV"     FOR
INPUT AS 1
      Input #ISW, SON
      Input #ISW, Sfile
      Input #ISW, Sname
```

```
    Input #ISW, SMAP
    Input #ISW, SDATE
    Input #ISW, Afido
    Input #ISW, Cfido
    Input #ISW, XD
    Input #ISW, YD
    Input #ISW, N
    For I = 0 To N
      Input #ISW, X(I), Y(I)
    Next I
' Terminate
    Close 1
    End Sub

    Sub DMEAN(N,NPP,ILB,IUB,ARY(),T,DMU)
' A Subroutine to Sum the Array Elements of ARY()
' Between Subscripts ILB and IUB inclusive
'
'  Arguments:
'     N        The Number of Intervals
'     NPP      The Number of Interval Boundaries
'     ILB      The Lower Bound
'     IUB      The Upper Bound
'     ARY()    Local Holder Array
'     T        Local Total
'     DMU      Local Mean
'
    Call DSUM(ILB, IUB, ARY(), T)
    DMU=T/NPP
' Terminate
    End Sub

    Sub DSUM (ILB, IUB, ARY(), T)
' A Function to Sum the Array Elements of ARY()
```

```
' Between Subscripts ILB and IUB inclusive
'
'  Arguments:
'     ILB        The Lower Bound
'     IUB        The Upper Bound
'     ARY()      Local Holder Array
'     T          Local Total
'
     Tot = 0.0
     For I = ILB To IUB
        Tot = Tot + ARY(I)
     Next I
     T=Tot
' Terminate
     End Sub

     Sub FILEOPEN (ISW, Spath, Sname, SFile, SExt)
' A Subroutine to Open a Sequential File for Input
'  Arguments:-
'     ISW        The Logical Unit Number
'     Spath      The Directories' Path
'     Sname      The Data Name
'     SFile      The Filename
'     SExt       The Filename Extension
'
     If ISW = 1 Then
          SExt=".CSV"
          SFile = "XYIN" + Sname
          Sconcat = Spath + SFile + SExt
          Open Sconcat For Input As ISW
       Else
          ISW=2
          SExt=".CSV"
     SFile = "SUMMARY" + Sname
```

```
        Sconcat = Spath + SFile + SExt
        Open Sconcat For Output As ISW
    End If
' Terminate
    End Sub

    Sub GRAVICENTER (N, XIN(), YIN(), XG, YG)
' A Subroutine to Compute the Gravitational Center of a Figure
' from its Cartesian Boundary Co-ordinates XIN() and YIN()
'
'  Arguments:
'    N         The Number of Sectors of Integration
'    XIN()     The Array of Input Cartesian X-values
'    YIN()     The Array of Input Cartesian Y-values
'    XG        The X-coordinate of the
'              Gravimetric Figure Center
'    YG        The Y-coordinate of the
'              Gravimetric Figure Center
' Zeroise Counters
    TX = 0
    TY = 0
' Total X and Y-displacements
    For I = 0 To N
        TX = TX + XIN(I)
        TY = TY + YIN(I)
    Next I
' Establish Cartesian Gravimetric Center
    XG = TX / N
    YG = TY / N
' Terminate
    End Sub

    Sub MEISTER (N, X(), Y(), AMe)
' A Subroutine to Compute the Meister's Rule
```

```
' Area from the Cartesian Co-ordinates X() and Y()
'
'  Arguments:-
'    N          The Number of ( Equal ) Intervals
'               of Array X()
'    X()        The Array of Values of x
'    Y()        The Array of Values of y
'    AMe        The Meister's Rule Area
'
     AMe = 0.0
     For I = 1 To N - 1
        AMe = AMe + Y(I) * (X(I - 1) - X(I + 1))
     Next I
     AMe=0.5*AMe
' Terminate
     End Sub

     Sub
OMNIBUS(N,X(),Y(),TH(),R(),XG,YG,RANGE,Afido,Asect,
Asimp,Cfido,Csegm,dth,fsindth,TH0,THN,h,Rmu,Acirc)
' A Subroutine to Compute the Characteristics of a Figure
'    Arguments:-
'        N       The Number of Radial Intervals
'        X()     The Array of Boundary x-Coordinates
'        Y()     The Array of Boundary y-Coordinates
'        TH()    The Array of Values of
'                Incremental Angles
'        R()     The Array of Values of Swept Radius
'        XG      The X-coordinate of the
'                Gravimetric Figure Center
'        YG      The Y-coordinate of the
'                Gravimetric Figure Center
'        RANGE The Total Swept Angle ( 2*PI radians )
'        Afido   The Fiducial Figure Area
```

```
'       Asect   The Sum of Sectors Figure Area
'       Asimp   The Simpson's Rule Figure Area
'       Cfido   The Fiducial Figure Perimeter
'       Csegm   The Figure Perimeter by Segment Summation
'
'       dth     The Increment of the Swept Angle
'       fsindth The Sine of the Increment of the Swept Angle
'       TH0     The Initial Sweep Angle
'       THN     The Final   Sweep Angle
'       h       The Simpson Increment
'       Rmu     The Mean Reference Radius
'       Acirc   The Reference Circle Area
'
' Establish Boundary Points
    dth=PI2/N
    fsindth=sin(dth)
    FOR I=0 to N
      TH(I)=I*dth
      R(I)=sqr(X(I)^2+Y(I)^2)
    NEXT I
    TH0=TH(0)
    THN=TH(N)
' Compute Area Parameters
    Tot1=0.0
    FOR I=0 TO N-1
      Tot1=Tot1+R(I)*R(I+1)*fsindth
    NEXT I
    Asect=0.5*Tot1
    Tot2=0.0
    FOR I=1 TO N-1
      Tot2=Tot2+(3+(-1)^(I-1))*R(I)^2
    NEXT I
    Asimp=(dth/6)*(R(0)+Tot2+R(N))
' Compute Perimeter Outputs
```

```
    Tot3=0.0
    FOR I=1 to N
      Tot3=Tot3+sqr((X(I)-X(I-1))^2+(Y(I)-Y(I-1))^2)
    NEXT I
    Csegm=Tot3
' Compute Auxiliary Variables
    RANGE=TH(N)-TH(0)
    h=RANGE/N
    Rmu=sqr(Afido/PI)
    Acirc=PI*Rmu^2
' Terminate
    End Sub

    Sub PRINTXY(N,X(),Y())
' A Subroutine to Print the Current Values of
' the N Data {X(I),Y(I)}
'   Arguments:-
'
'       N       The Number of Radial Intervals
'       X()     The Array of Boundary x-Coordinates
'       Y()     The Array of Boundary y-Coordinates
'
    I=0
    Ibatch=10
    DO WHILE I<N-1
       For K=1 to Ibatch
          I=I+1
          Print "X(";I;") = ";X(I);" Y(";I;") = ";Y(I)
      Next K
      Sleep
    LOOP
' Terminate
    End Sub
```

```
    Sub SIDES (N, X(), Y(), Csegm)
' A Subroutine to Estimate the Perimeter of an Ellipse by
' Summing its Side Segments
'  Arguments:-
'
'     N         The Number of Intervals of Array X()
'     X()       The Array of Values of x
'     Y()       The Array of Values of y
'     Csegm   The Sum of Side Segments
'
    Csegm = 0.0
    For I = 0 To N - 1
       DX = X(I + 1) - X(I)
       DY = Y(I + 1) - Y(I)
       Csegm = Csegm + Sqr(DX ^ 2 + DY ^ 2)
    Next
' Terminate
    End Sub

    Sub SIMPSONcartesian (N, RANGE, h,X(), Y(), ILB, IUB,
Asim)
' A Subroutine to Perform Simpson's Rule
' Numerical Integration upon the Array Y()
' of the Subject Abscissal Function
'
'  Arguments:-
'     N       The Number of ( Equal ) Intervals
'             of Array X()
'     RANGE   The Total Swept Angle ( 2*PI radians )
'     h       The Interval Width
'     X()     The Array of Values of x
'     Y()     The Array of Functions of x to be
'             Integrated
'     ILB     The Range Lower Bound
```

```
'     IUB    The Range Upper Bound
'     Asim  The Simpson Integral
'

      h = (X(N) - X(0)) / N
      TT = 0.0
      For I = 1 To N - 1
          TT = ALT(I) * Y()
      Next I
      Asim = (h / 3) * (Y(0) + TT + Y(N))
' Terminate
      End Sub

      Sub SIMPSONpolarType1 (N, RANGE, h, R(), TH(),
Asim)
' A Subroutine to Perform Simpson's Rule
' Numerical Integration upon the Array Y()
' of the Subject Abscissal Function.
' The Method is Adjusted for Polar Co-Ordinates.
'
'   Arguments:-
'      N              The Number of ( Equal ) Intervals
'                     of Array TH()
'      RANGE    The Total Swept Angle ( 2*PI radians )
'      h             The Interval Width
'      R()           The Array of Values of Swept Radius
'      TH()         The Array of Values of Incremental Angles
'      Asim        The Simpson Integral
'

      h=RANGE/N
      G0 = R(0) ^ 2
      GN = R(N) ^ 2
      TT = 0.0
      For I = 1 To N - 1
         TT = TT + ALT(I) * R(I) ^ 2
```

```
    Next I
    TT = G0 + TT + GN
    Asim = (1 / 6) * ((TH(N) - TH(0)) / N) * TT
' Terminate
    End Sub

    Sub SIMPSONpolarType2 (N, RANGE,h,R(), TH(), ILB,
IUB, Asim)
' A Subroutine to Perform Simpson's Rule
' Numerical Integration upon the Array Y()
' of the Subject Abscissal Function.
' The Method is Adjusted for Polar Co-Ordinates.
'
'  Arguments:-
'     N        The Number of ( Equal ) Intervals
'              of Array TH()
'     RANGE   The Total Swept Angle ( 2*PI radians )
'     h        The Interval Width
'     R()      The Array of Values of Swept Radius
'     TH()     The Array of Values of Incremental Angles
'     ILB      The Range Lower Bound
'     IUB      The Range Upper Bound
'     Asim     The Simpson Integral
'
    h = RANGE/N
    Dim G(1024)
    G(0) = R(0) ^ 2
    G(N) = R(N) ^ 2
    For I = 1 To N - 1
    G(I) = 0.0
    G(I) = R(I) ^ 2
    Next I
    Call SIMPSONcartesian(N, RANGE, h, TH(), G(), ILB,
IUB, Asim)
```

```
' Terminate
     End Sub

  Sub SUMMARYfigure (ISW,SON, SFile, Sname, SMAP,
SDATE,N, NPP, RANGE, XG, YG, XD, YD, dth, fsindth,TH0,
THN, h, Ameis, Afido, Asect, Asimp, Cfido, Csegm, Rmu,
Acirc)
' A Subroutine to Output the Summary Data and Results to
' File SUMMARYfigure.CSV on Logical Unit ISW=2
'
'  Arguments:-
'        ISW          The Logical Unit Number
'        SON          The Object Descriptor
'        Sfile        The Input File Name
'        Sname        The Data Name
'        SMAP         The Map Descriptor
'        SDATE        The Date of Data Compilation
'        N            The Number of Radial Intervals
'        NPP          The Number of Perimeter Points
'        RANGE        The Angular Range of the
'                     Integrative Sweep
         XG           The x-coordinate of the
'                     Gravimetric Figure Center
'        YG           The y-coordinate of the
'                     Gravimetric Figure Center
'        XD           The Desired Center x-Coordinate
'        YD           The Desired Center y-Coordinate
'        dth          Incremental Change in the Swept Angle
'        fsindth      The Sine of dth
'        TH0          The Initial Value of Sweep Angle
'        THN          The Final  Value of Sweep Angle
'        h            The Simpson Interval
'        Ameis        The Figure Area by Meister's Rule
'        Afido        The Fiducial Figure Area
```

```
'         Asect              The Figure Area by Sector Summation
'         Asimp              The Figure Area by the
'                            Polar Form of Simpson's Rule
'         Cfido              The Fiducial Figure Perimeter
'         Csegm              The Approximate Figure Perimeter
'         Rmu                The Radius of the Reference Circle
'         Acirc              The Area    of the Reference Circle
'

' Open the File
    SOUTpath = "D:\QUAD\BAS\"
    Sname = "SUMMARYellesmereN=80"
    SFile="Ellesmere"
    SExt = ".CSV"
    OPEN
"D:\QUAD\BAS\SUMMARYellesmereN=80.CSV"          FOR
OUTPUT AS 2
'   Call FILEOPEN(ISW, SOUTpath, Sname, SFile, SExt)
' Output the Data
    Write #ISW, "The Object Descriptor", SON
    Write #ISW, "The Input File Name", Sfile
    Write #ISW, "The Data Name", Sname
    Write #ISW, "The Map Descriptor", SMAP
    Write #ISW, "The Date of Data Compilation", SDATE
    Write #ISW,
    Write #ISW,"The x-coordinate of the Gravimetric Figure
Center",XG
    Write #ISW,"The x-coordinate of the Gravimetric Figure
Center",YG
    Write #ISW,"The Desired Center x-Coordinate",XD
    Write #ISW,"The Desired Center y-Coordinate",YD
    Write #ISW,
    Write #ISW, "The Number of Radial Intervals",  N
    Write #ISW, "The Number of Perimeter Points",  NPP
```

```
    Write #ISW, "The Angular Range of the Integrative
Sweep", RANGE
    Write #ISW, "Incremental Change in the Swept Angle", dth
    Write #ISW, "The Sine of dth", fsindth
    Write #ISW, "The Initial Value of Sweep Angle", TH0
    Write #ISW, "The Final  Value of Sweep Angle", THN
    Write #ISW, "The Simpson Interval", h
    Write #ISW,
    Write #ISW, "The Mean Radius", Rmu
    Write #ISW, "The Reference Circle Area", Acirc
    Write #ISW, "The Figure Area by Meister's Rule", AMeis
    Write #ISW,
    Write #ISW, "The Fiducial Figure Area", Afido
    Write #ISW, "The Figure Area by Sector Summation",
Asect
    Write #ISW, "The Figure Area by the Polar Form of
Simpson's Rule", Asimp
    Write #ISW,
    Write #ISW, "PSD(Afido,Asec)", PSD(Afido, Asect)
    Write #ISW, "PSD(Afido,Asim)", PSD(Afido, Asimp)
    Write #ISW, "PSD(Afido,AMe)", PSD(Afido, Ameis)
    Write #ISW, "PSD(Afido,Acirc)", PSD(Afido, Acirc)
    Write #ISW,
    Write #ISW, "The Fiducial Figure Perimeter", Cfido
    Write #ISW, "The Approximate Figure Perimeter", Csegm
    Write #ISW, "PSD(Cfido,Csegm)", PSD(Cfido, Csegm)
    Write #ISW,
    Radfido=Sqr(Afido/PI)
    AreaRefCir=PI*Radfido^2
    Write #ISW, "The Radius of the Reference Circle", Radfido
    Write #ISW, "The Area   of the Reference Circle",
AreaRefCir
    Write #ISW,
' Terminate
```

```
Close ISW
End Sub
```

Ellesmere (Figure) N=80 XYINEllesmere.CSV:- INPUT

Ellesmere
XYINEllesmere.DAT
Ellesmere
OS 1:25000 Great Britain 1937-1961
27 February 2026
460735
3000
905.0477958
899.9775
80
1143.351697,894.9356092
1128.14081,917.6241176
1120.535366,927.7078992
1107.859627,947.8754622
1102.789331,975.6058613
1100.254183,995.7734244
1100.254183,1018.461933
1107.859627,1043.671387
1102.789331,1071.401786
1092.648739,1091.569349
1085.043296,1106.695021
1064.762113,1126.862584
1052.086373,1144.509202
1036.875486,1162.155819
1021.664599,1187.365273
1008.988859,1220.137563
988.7076761,1237.784181
968.426493,1285.682143
935.4695704,1394.082794
897.4423521,1396.603739
864.4854296,1383.999013
823.9230634,1368.87334

798.5715845,1356.268613
760.5443662,1333.580105
732.6577394,1315.933487
699.7008169,1300.807815
669.2790423,1293.244979
636.3221197,1260.472689
623.6463803,1230.221345
610.9706408,1187.365273
608.435493,1147.030147
608.435493,1147.030147
605.9003451,1111.736912
598.2949014,1081.485567
583.0840141,1056.276113
565.3379789,1038.629496
534.9162042,1010.899097
507.0295775,990.7315336
504.4944296,963.0011345
512.0998732,930.2288445
527.3107606,899.9775
542.5216479,869.7261555
557.7325352,844.5167017
575.4785704,819.3072479
575.4785704,789.0559034
588.1543099,766.367395
605.9003451,743.6788866
623.6463803,723.5113235
636.3221197,708.3856513
659.1384507,683.1761975
676.8844859,665.5295798
699.7008169,652.9248529
709.8414085,640.3201261
745.3334789,632.7572899
768.1498099,625.1944538
788.430993,612.5897269

803.6418803,587.3802731
826.4582113,574.7755462
836.5988028,499.1471849
872.0908732,489.0634034
902.5126479,446.2073319
940.5398662,436.1235504
983.6373803,365.5370798
1036.875486,337.8066807
1077.437852,355.4532983
1082.508148,461.3330042
1115.46507,476.4586765
1115.46507,476.4586765
1163.63288,478.9796218
1206.730394,471.4167857
1244.757613,494.1052941
1277.714535,516.7938025
1297.995718,567.2127101
1305.601162,607.5478361
1287.855127,660.4876891
1259.9685,710.9065966
1206.730394,766.367395
1186.449211,804.1815756
1173.773472,829.3910294
1161.097732,862.1633193
1143.351697,894.9356092

Ellesmere (Figure) N=80 SUMMARYellesmereN=80.CSV:-OUTPUT

The Object Descriptor	Ellesmere
The Input File Name	Ellesmere
The Data Name	SUMMARYellesmereN=80
The Map Descriptor	OS 1:25000 Great Britain 1937-1961
The Date of Data Compilation	27-Feb-26
The x-coordinate of the Gravimetric Figure Center	897.5374201
The x-coordinate of the Gravimetric Figure Center	893.769672
The Desired Center x-Coordinate	905.0477958
The Desired Center y-Coordinate	899.9775
The Number of Radial Intervals	80
The Number of Perimeter Points	81
The Angular Range of the Integrative Sweep	6.283185307179590
Incremental Change in the Swept Angle	0.078539816339745
The Sine of dth	0.078459095727845
The Initial Value of Sweep Angle	0
The Final Value of Sweep Angle	6.283185307179590
The Simpson Interval	0.078539816339745

The Mean Radius	382.957576515844000
The Reference Circle Area	460735
The Figure Area by Meister's Rule	460769.717125545000000
The Fiducial Figure Area	460735.000000000000000
The Figure Area by Sector Summation	472066.088088381000000
The Figure Area by the Polar Form of Simpson's Rule	472216.642363450000000
PSD(Afido,Asec)	-2.459350404979310
PSD(Afido,Asim)	-2.492027383083640
PSD(Afido,AMe)	-0.007535161328151
PSD(Afido,Acirc)	-0.000000000000025
The Fiducial Figure Perimeter	3000.000000000000000
The Approximate Figure Perimeter	2912.278645548200000
PSD(Cfido,Csegm)	2.924045148393120
The Radius of the Reference Circle	382.957576500000000
The Area of the Reference Circle	460735

QUAD

WARREN

APPENDIX TWO
PARTIAL LISTS OF WIKIPEDIA ARTICLES

Please consult https://www.wikipedia.org/

Partial List of Wikipedia Articles abouts Lakes

Aqualate Mere
Bomere Pool
Brothers Water
Hatchmere
List of lakes of England
Llyn-y-Cau, Cader Idris
Loch Earn
Loch Enoch
Loch Ness
Rydal Water
Semerwater
Talkin Tarn

Partial List of Wikipedia Articles about Mathematics

Abramowitz and Stegun
Accuracy and precision
Area of a triangle
Cavalieri's principle
Complete Elliptic Integral of the Second Kind
Ellipse
Elliptic Integral
Fubini's Theorem
Gaussian quadrature
Lagrange Polynomial
Numerical Integration
Parabola
Quadratic Equation
Quadratrix
Quadrature (mathematics)
Quadrature of the Parabola
Simpson's Rule
Tanh-sinh quadrature
Universal parabolic constant
Verification and validation

QUAD

WARREN

www.ingramcontent.com/pod-product-compliance
Lightning Source LLC
LaVergne TN
LVHW051043210926
848486LV00001B/2